AF594991

AAAI Workshop on Artificial Intelligence with Biased or Scarce Data (AIBSD)

AAAI Workshop on Artificial Intelligence with Biased or Scarce Data (AIBSD)

Editors

Kuan-Chuan Peng
Ziyan Wu

MDPI • Basel • Beijing • Wuhan • Barcelona • Belgrade • Manchester • Tokyo • Cluj • Tianjin

Editors
Kuan-Chuan Peng
Mitsubishi Electric Research Laboratories (MERL)
USA

Ziyan Wu
United Imaging Intelligence
USA

Editorial Office
MDPI
St. Alban-Anlage 66
4052 Basel, Switzerland

This is a reprint of articles published online by the open access publisher MDPI (available at: https://www.mdpi.com/2813-0324/3/1). The responsibility for the book's title and preface lies with Kuan-Chuan Peng, who compiled this selection.

For citation purposes, cite each article independently as indicated on the article page online and as indicated below:

LastName, A.A.; LastName, B.B.; LastName, C.C. Article Title. *Journal Name* **Year**, *Volume Number*, Page Range.

ISBN 978-3-0365-4681-0 (Hbk)
ISBN 978-3-0365-4682-7 (PDF)

Cover image courtesy of Science Photo Library (Canva).

Contents

About the Editors

Kuan-Chuan Peng

Kuan-Chuan Peng, PhD is a Research Scientist at Mitsubishi Electric Research Labs (MERL) and a senior member of IEEE. He received his Ph.D. degree in Electrical and Computer Engineering from Cornell University in 2016. He co-organised the following events: (1) The Artificial Intelligence with Biased or Scarce Data Workshop in conjunction with AAAI in 2022; (2) The Fair, Data-Efficient and Trusted Computer Vision Workshop in conjunction with CVPR in 2020, 2021, and 2022; (3) The Vision Applications and Solutions to Biased or Scarce Data Workshop in conjunction with WACV 2020 and 2021; and (4) The Vision with Biased or Scarce Data workshop in conjunction with CVPR in 2018 and 2019. His relevant publications include guided and discriminative attention, zero-shot domain adaptation, etc.

Ziyan Wu

Ziyan Wu, PhD is a Principal Expert Scientist at UII America, Inc., and a senior member of IEEE. He received a Ph.D. degree in Computer and Systems Engineering from Rensselaer Polytechnic Institute, Troy, in 2014. He co-organized the following events: (1) The Artificial Intelligence with Biased or Scarce Data Workshop in conjunction with AAAI in 2022; (2) The Fair, Data-Efficient and Trusted Computer Vision Workshop in conjunction with CVPR in 2020, 2021, and 2022; (3) The Vision with Biased or Scarce Data workshop in conjunction with CVPR in 2018 and 2019; (4) The CVPR Industry EXPO Spotlight in 2017; and (5) The VIEW workshop in conjunction with CVPR in 2016. His relevant publications include interpreting with structural visual concepts, ConceptGAN, incremental scene synthesis, etc.

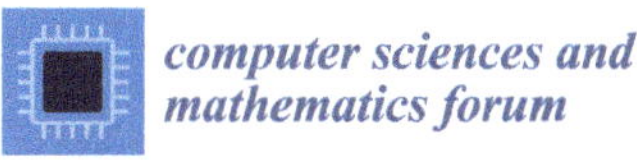

computer sciences and mathematics forum

MDPI

Editorial

Statement of Peer Review †

Kuan-Chuan Peng [1,*] and Ziyan Wu [2,*]

1 Mitsubishi Electric Research Laboratories (MERL), Cambridge, MA 02139-1955, USA
2 United Imaging Intelligence, Cambridge, MA 02140, USA
* Correspondence: kp388@cornell.edu (K.-C.P.); wuzy.buaa@gmail.com (Z.W.)

† All the papers are presented at the AAAI Workshop on Artificial Intelligence with Biased or Scarce Data (AIBSD), 28 February 2022; Available online: https://aibsdworkshop.github.io/2022.

In submitting conference proceedings to *Computer Sciences & Mathematics Forum*, the volume editors of the proceedings certify to the publisher that all papers published in this volume have been subjected to peer review administered by the volume editors. Reviews were conducted by expert referees to the professional and scientific standards expected of a proceedings journal.

- Type of peer review: double-blind
- Conference submission management system: CMT via https://cmt3.research.microsoft.com/AIBSD2022 (accessed on 12 November 2021).
- Number of submissions sent for review: 19
- Number of submissions accepted: 15 submissions are accepted to be presented at the AAAI Workshop on Artificial Intelligence with Biased or Scarce Data (AIBSD), and 10 of those 15 accepted submissions are encouraged to further submit to MDPI for publication.
- Acceptance rate (number of submissions accepted/number of submissions received): ~79% for the submissions presented at the AAAI Workshop on Artificial Intelligence with Biased or Scarce Data (AIBSD); ~53% for the submissions encouraged to further submit to MDPI for publication.
- Average number of reviews per paper: 2
- Total number of reviewers involved: 16
- Any additional information on the review process: The average number of papers assigned to each reviewer is only 2~3 to ensure the review quality.

Citation: Peng, K.-C.; Wu, Z. Statement of Peer Review. *CSFM* **2022**, 3, 12. https://doi.org/10.3390/cmsf2022003012

Published: 31 May 2022

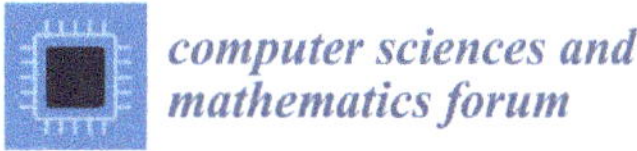
computer sciences and mathematics forum

MDPI

Proceeding Paper

Electricity Consumption Forecasting for Out-of-Distribution Time-of-Use Tariffs †

Jyoti Narwariya *,‡, Chetan Verma *,‡, Pankaj Malhotra, Lovekesh Vig, Easwara Subramanian and Sanjay Bhat

TCS Research, New Delhi 110 001, India; malhotra.pankaj@tcs.com (P.M.); lovekesh.vig@tcs.com (L.V.); easwar.subramanian@tcs.com (E.S.); sanjay.bhat@tcs.com (S.B.)

* Correspondence: jyoti.narwariya@tcs.com (J.N.); verma.chetan@tcs.com (C.V.)

† Presented at the AAAI Workshop on Artificial Intelligence with Biased or Scarce Data (AIBSD), Online, 28 February 2022.

‡ These authors contributed equally to this work.

Abstract: In electricity markets, electricity retailers or brokers want to maximize profits by allocating tariff profiles to end-consumers. One of the objectives of such demand response management is to incentivize the consumers to adjust their consumption so that the overall electricity procurement in the wholesale markets is minimized, e.g., it is desirable that consumers consume less during peak hours when the cost of procurement for brokers from wholesale markets are high. We consider a greedy solution to maximize the overall profit for brokers by optimal tariff profile allocation, i.e., allocate that tariff profile to a consumer that maximizes the profit with respect to that consumer. This, in turn, requires forecasting electricity consumption for each user for all tariff profiles. This forecasting problem is challenging compared to standard forecasting problems due to following reasons: (1) the number of possible combinations of hourly tariffs is high and retailers may not have considered all combinations in the past resulting in a biased set of tariff profiles tried in the past, i.e., the retailer may want to consider new tariff profiles that may achieve better profits; (2) the profiles allocated in the past to each user is typically based on certain policy, i.e., tariff profile allocation for historical electricity consumption data is biased. These reasons violate the standard IID assumptions as there is a need to evaluate new tariff profiles on existing customers and historical data is biased by the policies used in the past for tariff allocation. In this work, we consider several scenarios for forecasting and optimization under these conditions. We leverage the underlying structure of how consumers respond to variable tariff rates by comparing tariffs across hours and shifting loads, and propose suitable inductive biases in the design of deep neural network based architectures for forecasting under such scenarios. More specifically, we leverage attention mechanisms and permutation equivariant networks that allow desirable processing of tariff profiles to learn tariff representations that are insensitive to the biases in the data and still representative of the task. Through extensive empirical evaluation using the PowerTAC simulator, we show that the proposed approach significantly improves upon standard baselines that tend to overfit to the historical tariff profiles.

Keywords: out-of-distribution generalization; forecasting; temporal bias; permutation equivariance; optimization

Citation: Narwariya, J.; Verma, C.; Malhotra, P.; Vig, L.; Subramanian, E.; Bhat, S. Electricity Consumption Forecasting for Out-of-Distribution Time-of-Use Tariffs. *CSFM* **2022**, 3, 1. https://doi.org/10.3390/cmsf2022003001

Academic Editors: Kuan-Chuan Peng and Ziyan Wu

Published: 8 April 2022

1. Introduction

A smart grid consists of multiple types of entities such as those involved in generation, distribution, and consumption (smart appliances and buildings). One of the aims of a smart grid is to manage electricity demand in an economical manner via integration and exchange of information about all entities involved. For the customers or the end-consumers as well as the electricity distributing agencies or the electricity brokers, it offers the flexibility to choose/allocate among dynamically changing tariffs to meet certain objectives, e.g., minimize electricity bill for customers, maximize profit for retailers, etc. However, meeting such objectives is challenging due to dynamics of the market, e.g., changing wholesale electricity prices, supply–demand fluctuations, etc.

As depicted in Figure 1, a broker typically performs three functions: (1) purchase or sell power to its subscribers or customers in the retail market, (2) purchase or sell power in the wholesale market, and (3) rectify any supply–demand imbalance within its portfolio through the balancing market. In this work, we consider a simplified setting where the broker performs the following two functions: (1) sell power to those customers in the retail market who are electricity consumers, and (2) purchase power in the wholesale market. Typical examples of consumers include offices, housing complexes, hospitals, and villages. Furthermore, we focus on only those subset of consumers who have a *shiftable* load component in their total or aggregate consumption in addition to the traditional fixed or non-shiftable load, i.e., the consumption (e.g., appliance usage) at an hour that cannot be moved to another hour. This shiftable load can be shifted from the originally preferred hour to another hour in the day if the tariff for the latter is lower. The broker may want to encourage such a behavior, known as demand response management [1], to maximize profit or balance demand–supply.

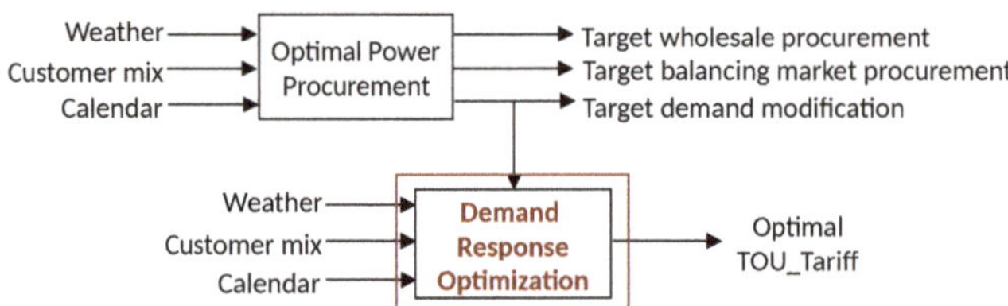

Figure 1. Various aspects and objectives in an electricity markets. In this work, we focus on a sub-problem related to allocation of optimal time-of-use tariff (TOU Tariff) to each customer.

In this work, we consider the following out-of-distribution generalization problem: given historical aggregated consumption of consumers to tariff profiles allocated to them, forecast the aggregated consumption for new tariff profiles. These new tariff profiles are part of the electricity broker or retailer's plan to explore new profiles to further improve the profits. This is different from standard forecasting problems as the exogenous variables (tariff profiles) at test time are different from the exogenous variables at train time. Furthermore, the allocation of tariff profiles in the past is not random, so the data is biased in the sense that, for different consumer personas, not all historical tariff profiles would have been tried. We note that the logic based on which the consumers respond to tariff profiles is consistent irrespective of the tariff profile. We propose to capture that logic in the neural network by using permutation equivariant networks and attention mechanisms.

The key contributions of this work can be summarized as follows:

- We consider the problem of electricity consumption forecasting under new tariff profiles not encountered previously. This is then used for tariff profile allocation to optimize electricity broker's profits.
- We note that the forecasting problem can be seen as an out-of-distribution (OOD) generalization problem with bias in the training data consisting of temporal and confounding bias.
- To achieve OOD generalization, we leverage the logic behind how consumers respond to tariff profiles in order to shift load, and propose a novel neural network architecture to achieve better OOD generalization.

Through empirical evaluation, we show that the proposed approach is able to improve upon vanilla methods that do not take into account suitable inductive biases guided by the knowledge of how consumers respond to tariff profiles.

2. Problem Formulation

The aggregated consumption $e_{c,t} \in \mathbb{R}^+$ of a consumer c at time t has two components: (1) *Type-I consumption*: this is non-shiftable consumption corresponding to the appliances that have to be used at specific hours only and cannot be shifted to alternative hour; (2) *Type-II consumption*: this is shiftable component of the consumption corresponding to appliances whose usage can be planned. Refer to Figure 2a for more details.

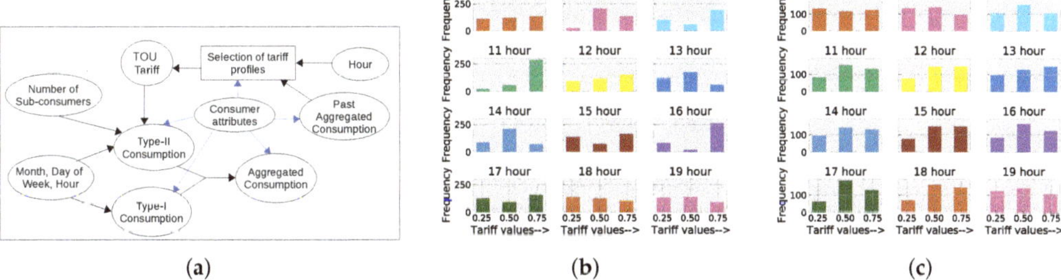

Figure 2. (**a**) Logic for Consumption Data generation in Electricity Markets and (**b**,**c**) Hourly Tariff Rate Distributions depicting changing distribution across hours that poses generalization challenge. (**a**) Causal Diagram. (**b**) Hourly Tariff Distributions in IID Profiles depicting temporal bias ($\mathcal{T}_{in}$). (**c**) Hourly Tariff Distributions in OOD Profiles ($\mathcal{T}_{out}$).

Let $e_{c,1:t}$ denote the time series of electricity consumption for consumer c until time t. We consider a consumer $c \in \mathcal{C}$, where $\mathcal{C}$ is the set of consumers with non-zero Type-II consumption, i.e., part of their load can be shifted in response to variations in tariff across hours. Further, the i-th time-of-use (TOU) tariff profile is denoted as an ordered sequence or H-length time series of hourly tariffs $TOU^i = TOU^i_1 \ldots TOU^i_H$, where TOU^i_h $(h = 1 \ldots H)$ denotes the tariff at hour h. In this work, we consider tariff profile with hourly rates over a day such that $H = 24$, without loss of generality.

Let $\mathbf{f}_{c,1:t}$ denote all features (static or time-varying) for consumer c at time t, including e.g., past consumption time series, type of consumer (household, office, etc.), and $\mathbf{f}_t$ denote a vector of temporal features at timestamp t, e.g., hour of the day, day of the week, week of the month, month of the year, etc. Note that $\mathbf{f}_{c,1:t}$ refers to relevant features from entire history, but in practice, we consider a window of length w over $t - w + 1 : t$ for deriving features at time t.

Further consider a tariff allocation policy function π such that

$$TOU_{c,t+\tau} = \pi(\mathbf{f}_{c,:t}, \mathbf{f}_{t+\tau}, \hat{p}_{t+1:t+H}),$$

i.e., the tariff at a future time $t + \tau$ with $\tau = 1 \ldots H$ is decided based on consumer features at time t, the temporal features for time $t + \tau$, where $\hat{p}_{t+\tau}$ denotes the estimate of electricity price $p_{t+\tau}$ in the wholesale market at time $t + \tau$. Without loss of generality, we consider the scenario where $t + 1$ corresponds to the first hour of the day, i.e., tariff profile for the next day is decided using data until the end of the current day.

Consider historical time series data $\mathcal{D} = \{e_{c,1:t}, TOU_{c,1:t}\}_{c \in \mathcal{C}}$, where the tariff time series are a result of sequence of tariff profile allocations over days such that any profile $TOU^i \in \mathcal{T}_{in}$ is chosen from a fixed set of profiles $\mathcal{T}_{in}$.

The goal for the broker is to allocate that tariff profile TOU^i to a consumer that maximizes the gain G^i_c over the next H hours:

$$G^i_c = \sum_{t'=1}^{H} (TOU^i_{c,t+t'} - p_{t+t'}) \times e_{c,t+t'}. \tag{1}$$

Importantly, the electricity consumption $e_{c,t+t'}$ at $t + t'$ hour is a function of the entire tariff profile on that day, as the consumer could choose to shift the shiftable part of the load from high tariff hours to low tariff hours by looking at the tariff profile allocated to the consumer at the beginning of the day.

We consider the following two scenarios depending on the tariff profiles being considered for future allocations:

IID Scenario: when the profiles to be allocated to the consumers in future are from the same set of profiles $\mathcal{T}_{in}$ used historically, i.e., $TOU^i \in \mathcal{T}_{in}$.

OOD Scenario: when the tariff profiles to be allocated to the consumers in future belong to $\mathcal{T}_{all} = \mathcal{T}_{in} \cup \mathcal{T}_{out}$, where $\mathcal{T}_{out}$ is a new set of profiles not previously seen in $\mathcal{D}$, i.e., are out-of-distribution with respect to the training data, and not previously allocated to any consumer by the broker who wants to consider these new profiles to improve future gains, i.e., $TOU^i \in \mathcal{T}_{all}$.

3. Related Work

Our work relates to two bodies of literature: (1) demand response management in electricity markets and the related sub-problem of electricity consumption forecasting under exogenous variables, using reinforcement learning and deep learning methods [2–4], and (2) out-of-distribution (OOD) generalization [5–8].

There have been many studies for (1); however, to the best of our knowledge, the problem of bias in historical data in terms of the tariff profiles has been largely overlooked. We draw attention of the community working on (1) to the potential of OOD generalization by improving forecasts for previously unallocated tariffs by using the underlying structure of the problem in terms of the particular way in which consumers shift loads in response to changes in tariff. More specifically, we rely on the partial permutation equivariance property of the response to time series of tariffs.

OOD detection and generalization is an emerging area of research, and aims at improving the robustness of models to previously unseen scenarios. Many of the recent approaches for (2) rely on changes in the objective function or different training procedures. For example, the approaches based on meta-learning [9] are not applicable as there is no notion of multiple tasks. We can consider each tariff profile as a task but then the forecasting can involve different profiles in input versus output. In this work, we focus on using inductive biases in the form of the neural network architecture to improve OOD generalization. There is enough evidence to support the improvement in generalization abilities of neural networks by using the structure of the problem to introduce suitable inductive biases in the learning process. The most commonly used inductive bias is in the design of the neural network architecture motivated by the structure of the problem. Recent examples of this include using graph neural networks [10,11] and modular networks [12]. Recently, using structural biases in deep neural networks motivated by the nature of bias and the structure of the problem have been successfully evaluated for time series forecasting [13]. Data-dependent priors have been recently proposed in [14]. However, to the best of our knowledge, using consumer behavior properties for electricity time series forecasting under out-of-distribution exogenous variables to guide the design of neural network architecture has not been considered so far in the literature.

4. The Learning Problem

We consider a 2-step approximate solution to maximize the gain (Equation (1)):

Step 1: For each consumer, forecast/estimate the consumption under each potential tariff profile allocation. Given features $\mathbf{f}_{c,1:t}$ (including $e_{c,1:t}$), history of allocated tariffs $TOU_{c,1:t}$, and values of potential future tariff $TOU_{c,t+1:t+H}$, the goal is to estimate $e_{c,t+1:t+H}$. This can be seen as **a multi-step time series forecasting problem with exogenous variables**. We provide the details of our proposed approach for this in the next section.

Step 2: Compute the profit using

$$\hat{G}_c^i = \sum_{t'=1}^{H} (TOU^i_{c,t+t'} - \hat{p}_{t+t'}) \times \hat{e}_{c,t+t'} \tag{2}$$

for each tariff in $TOU^i \in \mathcal{T}_{all}$ for OOD scenario ($\mathcal{T}_{in}$ for IID scenario). Allocate the tariff profile to consumer c which results in maximum $\hat{G}_c^i$. Note that, in practice, the future wholesale rates $p_{t+t'}$ ($t' = 1 \dots T$) are also not known and might need to be estimated. In this work, we assume that $p_{t+t'}$s are known in advance or estimable accurately and focus on estimating $\hat{e}_{c,t+t'}$s which are the only terms controllable via $TOU_{c,t+t'}$s.

In summary, the tariff profile allocation policy corresponds to estimating the gain for each tariff profile for a consumer, and then allocating the profile with maximum estimated gain. We use a deep neural network based architecture as the function approximator that estimates $\mathbb{E}[e_{c,t+t'}|TOU_{c,t+1:t+T}]$ from the data.

4.1. Biased and Scarce Data

The OOD scenario is challenging as there is no historical data for the profiles in $\mathcal{T}_{out}$. More concretely, we consider three possible values of tariff at any time t: low (0.2), medium (0.5), and high (0.8). Therefore, there are 3^H unique profiles possible. For $H = 24$, there can be $\approx 3 \times 10^{11}$ profiles possible. However, in practice, the number of allocated profiles would be significantly smaller than this. In this work, we consider $|\mathcal{T}_{in}| \in \{2, 5, 8, 10, 12, 15, 20, 30, 35\}$, which is a range of values encountered for $|\mathcal{T}_{in}|$ in practice. This poses serious OOD generalization challenge in estimating $e_{c,t+1:t+T}$ for previously unseen profiles $TOU^i_{t+1:t+T} \in \mathcal{T}_{out}$.

We note that one peculiar type of bias that manifests in practice is the **temporal bias**: at any hour h of the day, certain values of tariff are more common than others. We explain this further using a practical scenario as depicted in Figure 2: In practice, it is common to use the following heuristic for tariff profile allocation: Keep most expensive tariff rates during peak demand periods, least expensive tariff rates during non-peak hours, and slightly cheaper (medium) rates, typically between peak and off-peak periods. Every tariff profile is curated on the basis of average aggregated consumption of each customer. High tariff is allocated when the aggregated consumption is high, and for rest of the hours, low/mid tariff are allocated. The distribution of tariff rates over hours would depend on the distribution of peak consumption across customers (refer Figure 2c). Furthermore, there is confounding bias [15] with latent consumer attributes affecting (1) past aggregated consumption which in turn affects the treatment (tariff profile allocation), and (2) the outcome (electricity consumption) in $\mathcal{D}$ both can depend on the consumer features (refer Figure 2a). We leave the handling of confounding bias for future work, and focus on handling temporal bias in this work.

We empirically show that temporal bias poses a generalization challenge for vanilla feed-forward neural networks, and propose an attention-based architecture to deal with the same, in the next section.

4.2. How Consumers Respond to Tariffs

Consider the following toy example with $H = 6$ where there is only one tariff profile in $\mathcal{T}_{in}$ given by {*HHMMLL*}, i.e., tariff rate is high (H) for the first two hours, medium (M) for the next two hours, and low (L) for the last two hours. Further assume that the consumer has a certain Type-II load during the 1st hour. After looking at this tariff profile, the consumer responds by shifting the load from the 1st (high tariff) hour to the 5th (low tariff) hour. Now, consider a tariff profile in $\mathcal{T}_{out}$ as {*HHLLMM*}. Clearly, this profile is different from the profile in $\mathcal{T}_{in}$ as the sequence of highs and lows over the hours is different. However, importantly, the underlying decision-making behavior of the consumer remains the same, i.e., shift the Type-II load from high tariff hour (1st hour in this case) to low tariff hour (3rd hour instead of 5th hour in this case). Therefore, it is still possible to forecast the behavior of the user for this OOD profile. In this work, we intend to leverage this aspect of the consumer's decision-making process that stays the same irrespective of the IID-vs-OOD profiles.

Further, consider five ways to process the sequence of tariff rates (Figure 3):

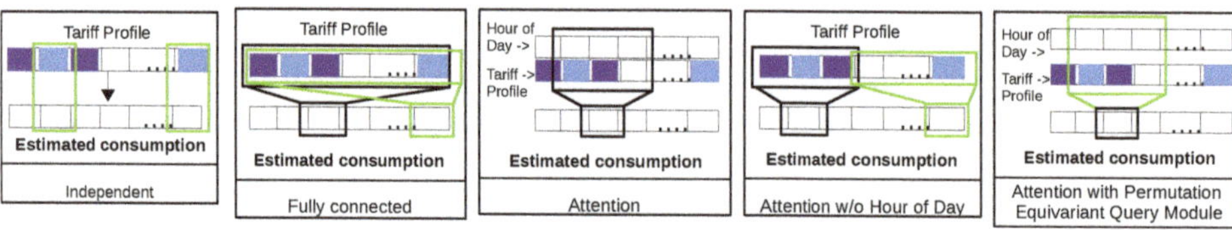

Figure 3. How different methods process the sequence of tariff rates.

- **Independent processing**: Here, the tariff at each hour is processed independently [16,17] and used to estimate the consumption at that hour. Of course, since the consumer's decision making is based on comparison of tariff rates across hours, such a processing of tariff profiles will not be effective.
- **All considered together or fully connected**: Here, tariffs at all hours (the entire tariff profile) are processed simultaneously, e.g., through a fully connected layer in a feed-forward neural network. We argue that such processing of tariff profiles will be able to effectively learn a good function approximator for the profiles in $\mathcal{T}_{in}$. However, it will be highly biased to the profiles in $\mathcal{T}_{in}$ since it does not effectively learn the way consumers are processing the tariff rates for shifting the loads. This leads to biased tariff profile processing modules due to the temporal bias in the historical profiles, as discussed above.
- **Focusing on relevant information or Attention**: Here, the tariffs rates in a day are considered as tokens and hours of a day are used as a positional information. This information is processed through a self-attention layer. We argue that such processing of tariff profiles will mimic the logic of how consumers respond to a tariff profile. However, it will be biased towards the profiles in $\mathcal{T}_{in}$ since the tariffs and hour of the day are correlated (due to temporal bias in the historical tariff profiles).
- **Permutation Equivariance**: As discussed earlier, permutation equivariance is an important aspect of the consumer decision-making logic. To mimic the same in the processing of tariffs by the neural networks, we expect that if trained on one of the tariff sequences, say, *HHMMLL* in the earlier example), it should perform equally well on other sequence (i.e., *HHLLMM*). In other words, processing of tariffs by neural networks should be Permutation Equivariant. We propose two ways to achieve approximate permutation equivariance:
  - **Attention w/o Hour of Day (Att.-HOD)**: As explained above, the standard self-attention method can mimic the logic of how consumers respond to tariffs, but due to temporal bias in the data, the attention method does not generalize well to $\mathcal{T}_{out}$. We propose a simple variant that does not take HOD as input in the self-attention module to obtain the permutation equivariance property.
  - **Attention with Permutation Equivariant Query Processing Module (Att.+PE)**: Here, the tariff rates in a day are considered as a set and processed in such a way that ordering of the tariff rates does not matter, i.e., the processing is permutation equivariant [18,19].

In the next section, we explain how we achieve permutation equivariance while forecasting the consumption given a consumer's consumption history, sequence of past tariff profiles, and a future tariff profile.

5. Forecasting Architecture

Consider the consumption history of a consumer along with past allocated tariffs to be a time series of vectors $\mathbf{f}_{1:t}$ including dimensions for past aggregate consumption and past allocated tariff rates $\{e_{1:t}, TOU_{1:t}\}$, and the candidate tariff profile for the next H hours to be $TOU_{t+1:t+H}$. The goal is to estimate $e_{t+1:t+H}$ while ensuring permutation equivariance in processing $TOU_{1:t+H}$ in the sense of [19], e.g., if the output of processing $\{TOU_1, TOU_2, TOU_3\}$ is $\{\mathbf{o}_1, \mathbf{o}_2, \mathbf{o}_3\}$, then the output of processing a permutation of the input, say $\{TOU_2, TOU_1, TOU_3\}$, is given by the permutation $\{\mathbf{o}_2, \mathbf{o}_1, \mathbf{o}_3\}$ of the original output.

To achieve the above-stated goal, we consider the following modularized neural network architecture as depicted in Figures 4 and 5:

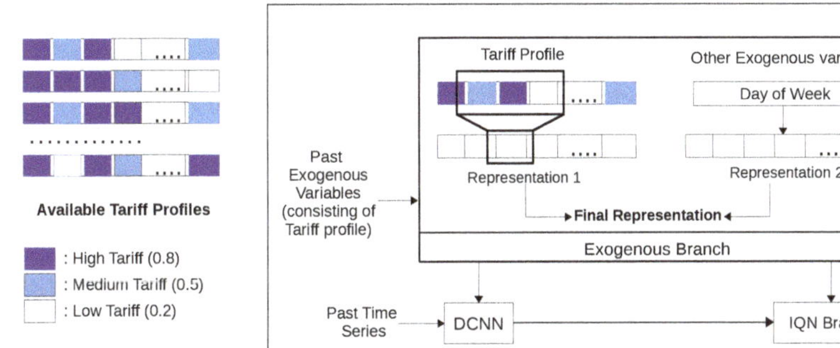

Figure 4. Flow diagram of "Attention w/o Hour of Day" approach. The left part of the figure indicates the variability in the tariff profiles and also some tariffs are more frequent in tariff profiles. The right part of the figure indicates flow of the inputs through the network and how the information of tariffs is consumed by the proposed approach.

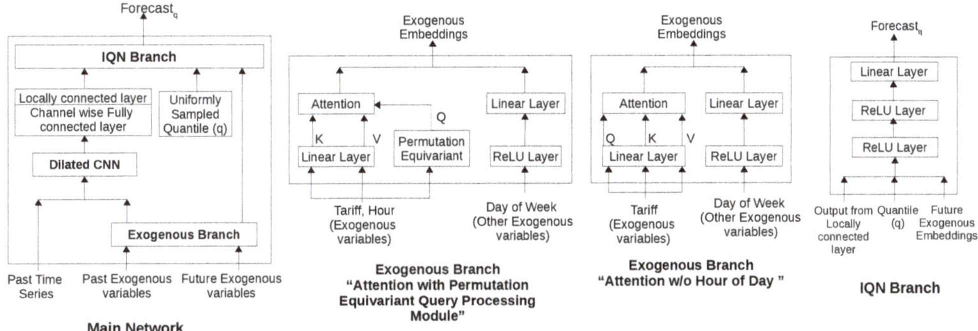

Figure 5. Architectures contrasting "Attention w/o Hour of Day" and "Attention with Permutation Equivariant Query Processing Module" approaches.

- **Dilated Convolutional Neural Networks (DCNN)** branch for processing of past consumption time series. (Since we have large input time series (t = 168 in our case), we consider 1D-Convolution Neural Networks for computational efficiency instead of Recurrent Neural Networks based architecture such as LSTMs [20].)
- **Exogenous** branch: This branch consists of **Attention with Permutation Equivariant Query Processing Module (Att.+PE)** branch for processing of tariff rates, and other modules for processing of features like hour of day, day of week, etc.
- **Implicit Quantile Network (IQN)** branch for generating the quantile estimates for future consumption.

Next, we provide details of the exogenous branch which is the key novel component of the proposed approach and helps to mitigate temporal bias.

To achieve permutation equivariance and handle temporal bias, we consider processing the tariff rates $TOU_{t+1:t+H}$ (same processing is done for past tariffs as well) via an attention mechanism where a part of the processing is done independently for tariff at each time step $t + t'$ ($t' = 1 \ldots H$) while still taking into account the global information

$TOU_{t+1:t+H}$ in order to mimic the behavior of the consumer as explained in the previous section.

More specifically, we consider key K and value V for the attention mechanism to be dependent on a single time step $t+t'$, while the query Q depends on the entire tariff profile $TOU_{t+1:t+H}$ for the day. In other words, $K_{t+t'} = f_K(TOU_{t+t'}, t+t', \theta_K)$, $V_{t+t'} = f_V(TOU_{t+t'}, t+t', \theta_V)$, and $Q_{t+t'} = f_Q(TOU_{t+1:t+H}, \theta_Q)$. Subsequently, the output for the part of the exogenous branch processing the tariffs at time $t+t'$ is given by

$$\mathtt{Att}(Q_{t+t'}, K_{t+t'}, V_{t+t'}) = \mathtt{softmax}(\frac{Q_{t+t'}K_{t+t'}^T}{\sqrt{d}})V_{t+t'}, \tag{3}$$

where d is the dimension of Q, K, and V. While the f_K and f_V are implemented as simple linear layers, f_Q is implemented as a permutation equivariant network as follows:

$$f(x) = \sigma(x\mathbf{\Lambda} - \mathbf{1}\mathtt{maxpool}(x)\mathbf{\Gamma}) \tag{4}$$

where $x = \mathtt{ReLU}(TOU_{t+1:t+H}, \theta_{TOU}) \in \mathbb{R}^{H\times d}$ and θ shared across timesteps $t+1\ldots t+H$, $\mathbf{\Lambda}, \mathbf{\Gamma} \in \mathbb{R}^{dxd'}$, matrix of ones $\mathbf{1} \in 1^{H\times H}$, `maxpool` is taken along columns implying that the resulting value for any timestep contains information from all timesteps and is independent of a particular timestep. In this work, we use $d = 10$, $d' = 20$.

Objective function: We use quantile loss for training the DCNN model given by:

$$\mathcal{L}_{quantile} = \frac{1}{b\times n}\sum_{i=1}^{b}\sum_{q=q_1}^{q_n} max(q\times e^i, (q-1)\times e^i), \tag{5}$$

where $e^i = y^i - \hat{y}^i$ indicates the error of the forecasted consumption $\hat{y}^i$ with respect to ground-truth consumption y^i of i-th window instance, b is the batch size and n is the number of quantiles used for training.

6. Experimental Evaluation

The goal is to evaluate the efficacy of the proposed approach to deal with OOD scenarios. For this, we compare the proposed approach with various baselines in the IID as well as OOD settings. We use the simulated data from a high-fidelity and popular PowerTAC (https://powertac.org/, accessed on 12 November 2021) [21] simulator that uses complex state-of-the-art user-behavior models and real world weather data to simulate the complex dynamics of a smart grid system.

We consider 'Office Complex Controllable type' consumers where consumers' daily behavior depends on factors such as number of sub-customers, number of appliances, weather information, hour of day, month, day of week, etc. The various values these factors can take across consumers is given in Table 1.

Table 1. Dataset details.

S.N.	Properties of Consumers	Value(s)
1	Number of consumers	12
2	Number of sub-consumers	3, 5
3	Working days	3, 4
4	Work Start hour	{8, 9, 10} (+/−) 1 h
5	Break Start hour	{13, 14} (+/−) 1 h
6	Work duration	8 (+/−) 1 h
7	Shiftable consumption(in KW)	600, 2400
8	Total data duration (in months)	6

To obtain train, validation, and test split, we divide the total data of 6 months into 4, 1, and 1 month, respectively. The time series of hourly data for each consumer is divided into windows of length $t = 168$ (corresponding to 7 days) with window-shift of 24 to forecast one day-head consumption, i.e., output window size is 24. We consider varying number of tariff profiles in historical data, i.e., $|\mathcal{T}_{in}| \in \{2, 5, 8, 10, 12, 15, 20, 25, 30, 35\}$, and an additional set of $|\mathcal{T}_{out}| = 40$ profiles. As the number of profiles $|\mathcal{T}_{in}|$ in the training set increases, we expect the bias in the training data to reduce.

6.1. Baselines Considered

For comparison, we consider the following approaches all using DCNN as the core time series processing module:

- **No future exogenous variable (NoX)** is the simple univariate time series forecasting approach which uses only history of aggregated consumption without any additional future information. This can be considered as a lower bound in the sense that the network does not have access to any future tariff rates to estimate where a consumer will shift the load.
- **Independent tariff-based method (Ind.)** is an approach that treats each tariff rate independently, and uses the tariff at time $t + t'$ to estimate the aggregated consumption at that time. Importantly, this approach has no means to capture comparison of the tariff rates in order to figure out whether the tariff at time $t + t'$ is high or low in comparison to another timestep.
- **Fully-Connected Approach (FC)** utilizes the information of all timesteps to estimate the aggregated consumption at each timestep. As explained previously, we expect such an approach to perform well in the IID scenario but struggle in the OOD scenario where new profiles are included.
- **Permutation Equivariant (PE)** method uses only the permutation equivariance idea from our approach and ignores the attention mechanism. This method can be thought of as an ablation over our approach.
- **Attention (Att.)**: This is another ablation over our approach which uses standard attention module for processing the tariffs along with hour of the day information without any permutation equivariance property.
- **Upper Bound (UB)**: This is an oracle approach that assumes knowledge about the hours at which the consumer is going to shift the load. In this, a binary value indicating whether the shiftable load will be shifted to this hour or not is passed as an additional feature to the exogenous branch of the Att.+PE network.

6.2. Hyperparameters Used

We use z-normalized consumption time series. DCNN has three layers with each layer having 16 convolutional filters of length 2, and dilation rate 1, 2, and 4, respectively. We use batch normalization and L2 filter regularizer ($\lambda = 0.001$) for regularization purposes. ReLU layers are applied on each CNN layer. The output of the DCNN layer is processed by a channel-wise fully connected layer, which has 24 hidden units (equal to the output window size) i.e., 24, followed by locally connected layer with 10 filters which are applied at each time-step independently (filter size = 1).

To obtain categorical feature (hour of day, day of week, month of year) embeddings and tariff rate embeddings, we use a separate feed-forward network with ReLU layer followed by linear layer, having 5 hidden units and 10 hidden units respectively. Similarly, we use 10 hidden units for each feed-forward network f_Q, f_K, f_V. Finally, the output layer is a small feed-forward network that has 2 layers followed by a linear layer having 40, 10, and 1 hidden unit, respectively. We use batch size of 16, number of epochs 200, and Adam optimizer with fixed learning rate of 0.0001 for training the neural network. During training, quantiles are sampled from uniform distribution while during validation and testing, we use three quantiles 0.1, 0.5, and 0.9. All hyperparameters were obtained via grid search based on validation quantile loss on the IID set.

6.3. Results and Observations

We make following key observations from the results in Figures 6 and 7:

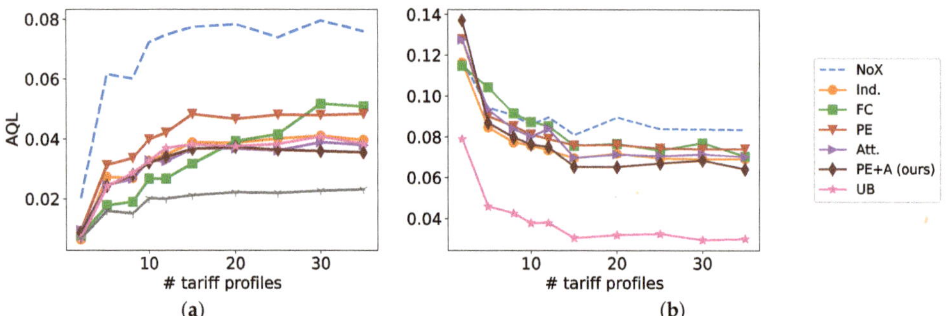

Figure 6. Forecasting performance Comparison of different approaches (in terms of Average Quantile Loss). (**a**) IID Scenario. (**b**) OOD Scenario.

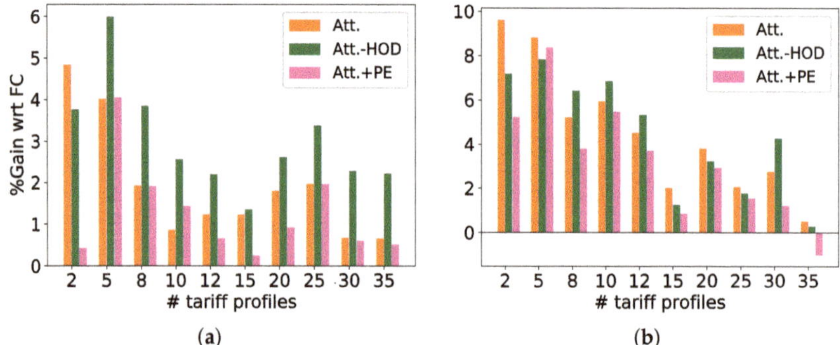

Figure 7. %gains of the proposed Att.+PE, Att.-HOD, and Att. approaches over the vanilla FC approach. (**a**) Option-1. (**b**) Option-2.

- Observations from forecasting results as shown in Figure 6:
  - In the IID scenario, the average quantile loss (AQL) for all approaches increases with increasing number of tariff profiles as the complexity of the dataset increases. The FC approach performs better than other approaches for $|\mathcal{T}_{in}| \leq 15$, indicating higher expressivity of the FC approach to fit to a smaller number of IID profiles, indicating potential overfitting.
  - On the other hand, for the OOD scenario, the performance of all approaches improves with increasing number of IID profiles which is expected as more IID profiles implies less bias and better generalization to OOD profiles as well. Interestingly, the FC approach which was the best approach for the IID profiles for $|\mathcal{T}_{in}| \leq 12$, is the worst approach (except the lower bound NoX) in the OOD setting, because it uses a fully connected layer to process the tariffs of the day, and due to temporal bias in the data, the weights of fully connected layer will try to overfit on $|\mathcal{T}_{in}|$ and thus not generalize to OOD profiles$|\mathcal{T}_{out}|$.

    On the other hand, our proposed approaches Att.+PE and Att.-HOD are consistently better than FC for all values of $\mathcal{T}_{in}$, which shows that FC struggles with the temporal bias in the historical data. We also analyze that Att.-HOD as well as Att.+PE are also consistently better than Att. for all values of $\mathcal{T}_{in}$, which shows that permutation equivariant way of handling tariff profiles provide better generalization on OOD profiles.

- We further analyze whether the gains of Att.+PE and Att.-HOD over other methods on the OOD scenario translate into more profitable tariff profile allocation for the retailer. We compare the gain G of Att.+PE, Att.-HOD, and Att. in comparison to FC. We consider two kinds of profiles for wholesale prices p, one with two values (0.2 and 0.8, referred to as Option-1) and one with three values (0.2, 0.5, and 0.8, referred to as Option-2).
  - **Comparison with FC**: We observe that all attention-based proposed approaches Att., Att.-HOD, and Att.+PE depict significant positive gains over FC. We also observe that Att., Att.-HOD, and Att.+PE approaches have higher positive gain in fewer IID tariff profiles scenarios $|\mathcal{T}_{in}| \leq 12$ (except $|\mathcal{T}_{in}| = 2$, where data is too little to claim any generalization), and the gains tend to diminish as $|\mathcal{T}_{in}|$ increases.
  - As expected, we note that it is not important that the gains in forecasting translate directly into monetary profits, as the optimization objective involves other terms such as wholesale costs p. Therefore, the best approach on forecasting (Att.+PE) in the OOD scenario is not necessarily the best approach in terms of profit always.
  - **Comparison with Att.**: For Option-1, Att.-HOD has significantly better gains than Att. for all values of $\mathcal{T}_{in}$ except $|\mathcal{T}_{in}| = 2$, which shows that the permutation equivariant way of handling tariff profiles is helpful. For Option-2, the gains of Att.-HOD are better or close to the gains of Att. approach (except $|\mathcal{T}_{in}| = 2$).

In Figure 8, we also provide sample forecasts comparing Att., Att.-HOD, Att.+PE, and FC with the ground truth (GT) on an OOD profile, indicating better generalization ability of Att.-HOD and Att.+PE, especially around points where Type-II load gets shifted. On the other hand, all methods perform well in the IID setting as shown in Figure 9.

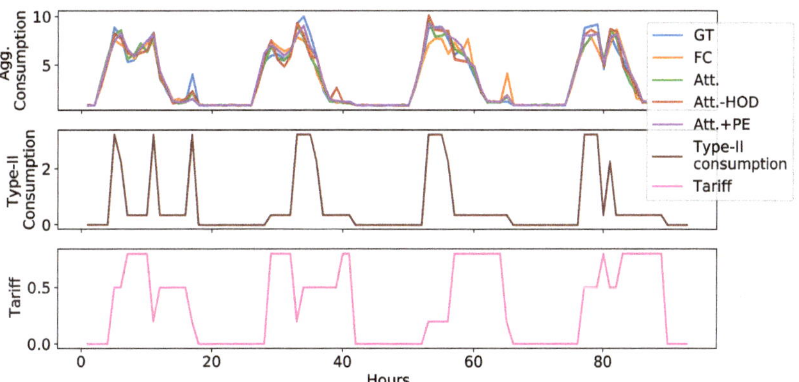

Figure 8. Sample results comparing the proposed approaches Att.-HOD and Att.+PE with FC on an OOD tariff profile. Here, GT: Ground Truth time series. FC struggles to capture the subtle changes in consumption due to shifting of load, while both Att.-HOD and Att.+PE are able to forecast better.

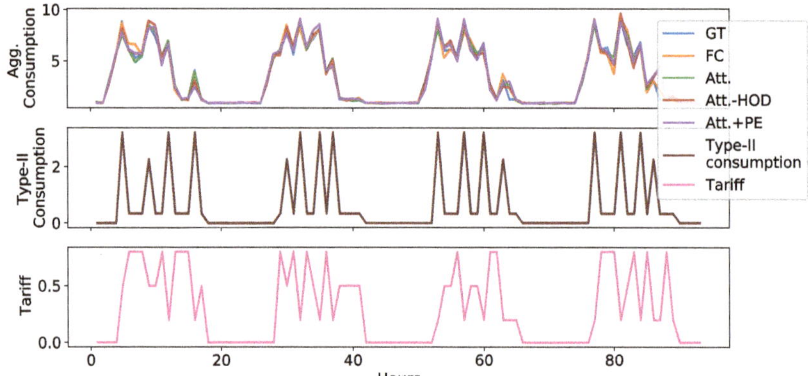

Figure 9. Sample results comparing the proposed approaches Att.-HOD and Att.+PE with FC on an IID tariff profile. Here, GT: Ground Truth time series. In IID scenario, all proposed attention-based approaches and baselines perform well.

7. Conclusions and Future Work

In this work, we consider the problem of demand response management from an electricity broker or retailer's perspective. We highlight temporal bias as an issue in optimizing profits via suitable tariff profile allocations. We motivate the need for better generalization to out-of-distribution profiles, and note that this is possible by leveraging the fact that consumers respond with same logic across profiles. We propose suitable inductive biases in deep neural networks-based approach for forecasting electricity consumption in response to new tariff profiles. This takes the form of a permutation equivariance-enabled attention mechanism that can leverage the property of consumer behavior to respond in a certain way across profiles. In the future, it will be interesting to look at the generalization from the perspective of handling confounding bias as the historical profile allocation and the outcome are affected by the historical allocation policies, which in turn rely on the latent consumer attributes acting as confounders. The current optimization objective takes into account broker's profit but ignores the cost of electricity for the end consumer—bringing this into the optimization objective is a potential next step.

Author Contributions: Conceptualization, methodology, resources, software, formal analysis and writing of original draft, P.M., J.N. and C.V.; Validation and data curation, J.N. and C.V.; writing—editing and review L.V., E.S. and S.B.; supervision, P.M., L.V., E.S. and S.B. All authors have read and agreed to the published version of the manuscript.

Funding: This research received no external funding.

Institutional Review Board Statement: Not applicable.

Informed Consent Statement: Not applicable.

Data Availability Statement: We use simulated data from PowerTAC simulator. Further details about the data are provided in Section 6. Data is confidential, so we can not provide the simulated data.

Conflicts of Interest: The authors declare no conflict of interest.

References

1. Siano, P. Demand response and smart grids—A survey. *Renew. Sustain. Energy Rev.* **2014**, *30*, 461–478. [CrossRef]
2. Lu, R.; Hong, S.H. Incentive-based demand response for smart grid with reinforcement learning and deep neural network. *Appl. Energy* **2019**, *236*, 937–949. [CrossRef]
3. Lu, R.; Hong, S.H.; Zhang, X. A dynamic pricing demand response algorithm for smart grid: Reinforcement learning approach. *Appl. Energy* **2018**, *220*, 220–230. [CrossRef]
4. Yang, P.; Tang, G.; Nehorai, A. A game-theoretic approach for optimal time-of-use electricity pricing. *IEEE Trans. Power Syst.* **2012**, *28*, 884–892. [CrossRef]

5. Hendrycks, D.; Basart, S.; Mu, N.; Kadavath, S.; Wang, F.; Dorundo, E.; Desai, R.; Zhu, T.; Parajuli, S.; Guo, M.; et al. The many faces of robustness: A critical analysis of out-of-distribution generalization. In Proceedings of the IEEE/CVF International Conference on Computer Vision, Montreal, BC, Canada, 11–17 October 2021; pp. 8340–8349.
6. Arjovsky, M. Out of Distribution Generalization in Machine Learning. Ph.D. Thesis, New York University, New York, NY, USA, 2020.
7. Krueger, D.; Caballero, E.; Jacobsen, J.H.; Zhang, A.; Binas, J.; Zhang, D.; Le Priol, R.; Courville, A. Out-of-distribution generalization via risk extrapolation (rex). In Proceedings of the International Conference on Machine Learning. PMLR, Virtual Event, Switzerland, 7–8 June 2021; pp. 5815–5826.
8. Sun, Y.; Wang, X.; Liu, Z.; Miller, J.; Efros, A.A.; Hardt, M. Test-time training for out-of-distribution generalization. In Proceedings of the Eighth International Conference on Learning Representations (ICLR 2020), Virtual Conference, 26 April–1 May 2020.
9. Hospedales, T.; Antoniou, A.; Micaelli, P.; Storkey, A. Meta-learning in neural networks: A survey. *arXiv* **2020**, arXiv:2004.05439.
10. Wang, T.; Liao, R.; Ba, J.; Fidler, S. Nervenet: Learning structured policy with graph neural networks. In Proceedings of the International Conference on Learning Representations, Vancouver, BC, Canada, 30 April–3 May 2018.
11. Narwariya, J.; Malhotra, P.; TV, V.; Vig, L.; Shroff, G. Graph Neural Networks for Leveraging Industrial Equipment Structure: An application to Remaining Useful Life Estimation. *arXiv* **2020**, arXiv:2006.16556.
12. Andreas, J.; Rohrbach, M.; Darrell, T.; Klein, D. Neural module networks. In Proceedings of the IEEE Conference on Computer Vision and Pattern Recognition, Honolulu, HI, USA, 21–26 July 2016; pp. 39–48.
13. Bansal, H.; Bhatt, G.; Malhotra, P.; Prathosh, A. Systematic Generalization in Neural Networks-based Multivariate Time Series Forecasting Models. In Proceedings of the 2021 International Joint Conference on Neural Networks (IJCNN), Shenzhen, China, 18–22 July 2021; pp. 1–8.
14. Liu, T.; Lu, J.; Yan, Z.; Zhang, G. Statistical generalization performance guarantee for meta-learning with data dependent prior. *Neurocomputing* **2021**, *465*, 391–405. [CrossRef]
15. Pearl, J.; Glymour, M.; Jewell, N.P. *Causal Inference in Statistics: A Primer*; John Wiley & Sons: Hoboken, NJ, USA, 2016.
16. Salinas, D.; Flunkert, V.; Gasthaus, J.; Januschowski, T. DeepAR: Probabilistic forecasting with autoregressive recurrent networks. *Int. J. Forecast.* **2020**, *36*, 1181–1191. [CrossRef]
17. Liu, X.; Yin, J.; Liu, H.; Liu, J. DeepSSM: Deep State-Space Model for 3D Human Motion Prediction. *arXiv* **2020**, arXiv:2005.12155.
18. Zaheer, M.; Kottur, S.; Ravanbakhsh, S.; Poczos, B.; Salakhutdinov, R.; Smola, A. Deep sets. *arXiv* **2017**, arXiv:1703.06114.
19. Lee, J.; Lee, Y.; Kim, J.; Kosiorek, A.; Choi, S.; Teh, Y.W. Set transformer: A framework for attention-based permutation-invariant neural networks. In Proceedings of the International Conference on Machine Learning. PMLR, Long Beach, CA, USA, 9–15 June 2019; pp. 3744–3753.
20. Hochreiter, S.; Schmidhuber, J. Long short-term memory. *Neural Comput.* **1997**, *9*, 1735–1780. [CrossRef] [PubMed]
21. Ketter, W.; Collins, J.; Reddy, P. Power TAC: A competitive economic simulation of the smart grid. *Energy Econ.* **2013**, *39*, 262–270. [CrossRef]

computer sciences and mathematics forum

MDPI

Proceeding Paper

Measuring Embedded Human-Like Biases in Face Recognition Models †

SangEun Lee [1,‡], Soyoung Oh [1,‡], Minji Kim [1] and Eunil Park [1,2,*]

[1] Department of Applied Artificial Intelligence, Sungkyunkwan University, Seoul 03063, Korea; sange1104@g.skku.edu (S.L.); sori424@g.skku.edu (S.O.); m5512m@g.skku.edu (M.K.)
[2] AI Team, Raon Data, Seoul 04522, Korea
* Correspondence: eunilpark@skku.edu
† Presented at the AAAI Workshop on Artificial Intelligence with Biased or Scarce Data (AIBSD), Online, 28 February 2022.
‡ These authors contributed equally to this work.

Abstract: Recent works in machine learning have focused on understanding and mitigating bias in data and algorithms. Because the pre-trained models are trained on large real-world data, they are known to learn implicit biases in a way that humans unconsciously constructed for a long time. However, there has been little discussion about social biases with pre-trained face recognition models. Thus, this study investigates the robustness of the models against racial, gender, age, and an intersectional bias. We also present the racial bias with a different ethnicity other than white and black: Asian. In detail, we introduce the Face Embedding Association Test (FEAT) to measure the social biases in image vectors of faces with different race, gender, and age. It measures social bias in the face recognition models under the hypothesis that a specific group is more likely to be associated with a particular attribute in a biased manner. The presence of these biases within DeepFace, DeepID, VGGFace, FaceNet, OpenFace, and ArcFace critically mitigate the fairness in our society.

Keywords: face-recognition models; facial attributes; social bias; fairness

Citation: Lee, S.; Oh, S.; Kim, M.; Park, E. Measuring Embedded Human-Like Biases in Face Recognition Models. *CSFM* **2022**, 3, 2. https://doi.org/10.3390/cmsf2022003002

Academic Editors: Kuan-Chuan Peng and Ziyan Wu

Published: 11 April 2022

1. Introduction

Recent advances in machine learning technologies allow computer vision researchers to employ massive datasets from the web to train models with image representations for general purposes from face recognition to image classification [1,2]. However, the absence of scrutinizing those datasets disproportionately can cause negative impacts on racial and ethnic minorities as well as other vulnerable individuals [3]. Without the necessary precautions of these problematic narratives, there can be some issues in image classification and labeling practices that entail stereotypes and prejudices [4,5]. The machine learning models with such datasets may elaborate and normalize these stereotypes, inflicting unprecedented harm on those who already comprise the margins of our society.

Therefore, it is essential to understand how datasets are sourced, labeled, and what representations the models are trained on. One of the common measures called the Word Embedding Association Test (WEAT) is used to assess undesirable associations in word embeddings [6]. That is, WEAT is used to show that both humans and natural language processing reveal many of the same biases with similar significance. For instance, WEAT shows racial bias in the word vector space by quantifying the close relations between pleasant words and European American names and unpleasant words with African American names. Ross et al. [7] extend this work with a metric throughout interaction between vision and language embeddings to measure biases in social and cultural concepts, such as race. We extend prior works with a metric, which we term Face Embedding Association Test (FEAT) to probe race, gender, and age biases in embeddings of pre-trained face recognition models. Unlike the previous measurements that measure bias within the facial image representation itself [8,9], our measurement measures evaluative associations between pairs of semantic categories which resemble the implicit attitudes underlying human cognitive priming procedure [10]. That is, FEAT measures the models' automatic associations as if estimating humans' stereotypical discrimination toward social categories represented by associations between a target and an attribute dimension. In addition, a strong advantage of FEAT is its potential for extension to additional discrimination tests. It is adaptable to assess a wide range of biases in our society.

By taking advantage of the expandability of FEAT, we expand to assess social biases toward a relatively unexplored racial group. There have been a lack of studies measuring biases of various races but only focused on white and black ethnicity. It is a significant oversight to invalidate ethnic group differences within racial category, which is another common form of discrimination experienced not only by Asian people but by other racial groups as well [11]. Understanding nuances in how different groups of people are affected by their ethnicities represents the next step in advancing this field of study. Thus, we take the next step to answer the question whether the models are significantly affected by the biases toward other racial groups rather than white and black. To achieve this goal, we employ face images of European American (EU), African American (AF), and Asian American (AS) people. Moreover, we measure an interaction between racial and gender biases that submissiveness and incapable of becoming leaders is prevalent in Asian women [12]. In short, our contributions are:

- We introduce FEAT to measure racial, gender, age, and an intersectional bias in face recognition models with images.
- We find statistically significant social biases embedded in pre-trained DeepFace [13], DeepID [14], VGGFace [15], FaceNet [16], OpenFace [17], and ArcFace [18].
- Our new dataset and implementations are publicly available (https://github.com/sange1104/face-embedding-association-test, accessed on 28 February 2022).

2. Related Work

A bias mitigation method can be largely divided according to the areas of model distribution targeted for pre-processing, in-processing, and post-processing [19]. The most widely used pre-processing technique is to re-balance datasets [20,21] or use synthetic data [22]. In the case of datasets used in face recognition tasks, they proved to have an imbalanced class distribution both in gender and race [23]. To address this problem, several datasets with a balanced number of gender, ethnicity, and the other attributes are proposed by the previous studies, including Racial Faces in Wild [24], Balanced Faces in the Wild [25], and DiveFace [26]. Although, these datasets contribute to mitigating abnormal distributions, but not to demonstrating that training with these datasets leads to impartial results, because labels for ethnicity in the datasets are not widely allowed as ground truth and are overly dependent on the annotator's decision [27]. This motivates researchers to develop in- and post-processing methods.

In-processing approaches take several methods to get rid of impartiality while training. For example, cost-sensitive training and adversarial learning are used to get rid of sensitive information from functionality [20,21]. Moreover, adjusting parameters of loss functions and taking an unsupervised way of training are used to protect minorities by training models with unbiased representations [26,28]. The examples of post-processing techniques include re-regulating the similarity scores of the two feature vectors based on demographic groups of the images [29] or attaching layers to the feature extractor for removing sensitive information from the representation [26].

Along the line, growing numbers of measurements have appeared to measure the effectiveness of the mitigation approaches. In the natural language processing field, various tests have been proposed to quantify bias in pre-trained word embedding models. Bolukbasi et al. [30] and Manzini et al. [31] employed word analogy tests and demonstrated undesirable bias toward gender, racial, and religious groups in word embeddings.

Moreover, Nadeem et al. [32] present a new evaluation metric that measures how close a model is to an idealistic model, showing that word embeddings contain several stereotypical biases.

Though less work has been studied to measure bias in the computer vision area compared to text, there are several approaches to examine embedded bias in visual recognition tasks. Acien et al. [33] investigate to what extent sensitive data such as gender or ethnic origin attributes are present in the face recognition models. Wang et al. [34] propose a set of measurements of the encoded bias in vision tasks and demonstrate that models amplify the gender biases with an existing dataset. Furthermore, recent studies focus on generation models to explore biases in face classification systems [22,35].

One of the widely used methods to examine bias is evaluating the representation produced by the model [6,36], as it can be easily utilized as a tool to analyze human bias [37,38]. To analyze the implicit bias, the WEAT [6] calculates word associations between target words and attribute words. Replacing words to sentences, the Sentence Encoder Association Test (SEAT) is introduced to apply WEAT to measure biases in sentence embeddings [39]. Moreover, recent studies generalize WEAT to contextualized word embeddings and investigate gender bias in contextual word embeddings from ELMo [40,41]. Steed and Caliskan [1] adapt WEAT to the image domain to evaluate embedded social biases. However, to our knowledge, there are no principle tests for measuring bias toward diverse racial subgroups, especially for Asians with face recognition models. Our work aims to generalize WEAT to facial image embeddings in order to examine social biases toward a wide range of subgroups in pre-trained face recognition models.

3. Methods

3.1. Face Embedding Association Test

Existing bias measures in natural language processing assess bias of word or sentence based on an Implicit Association Test administered to humans [6,42,43]. We introduce Face Embedding Association Test (FEAT) by extending the prior works throughout face embeddings. The details of the FEAT are as follows.

FEAT uses sets of face images, rather than sets of words or sentences, to demonstrate race and gender. Two sets of face images, X and Y, denote two sets of target races of the same size, while A and B are two sets of attribute images. For example, as in Figure 1, a face image x represents EU, while y as AS. One example of career attribute images A denote as a and b is an example of family attributes B. The basis of an indicator of bias is calculated by the average cosine similarity between pairs of images. Equation (1) measures the association of one of the target face images f with different attributes as follows:

$$s(f, A, B) = mean_{a \in A} cos(f, a) - mean_{b \in B} cos(f, b) \tag{1}$$

where the s function measures how close an average embedding for face image f with attribute set A compared to the B. The relative proximity of f and A opposed to B indicates that both concepts are more closely related.

Then, all target face images (i.e., X and Y) can be used to measure the bias in vector space. Bias is defined as one of the two target sets being significantly closer to one set of attribute images compared to the other. For example, the social bias is present when it comes to one of the target sets EU or AS is significantly closer to the concept of career compared to family. The following equation, $s(X, Y, A, B)$, measures the differential association of the two sets of target images with the attribute:

$$s(X, Y, A, B) = \sum_{x \in X} s(x, A, B) - \sum_{y \in Y} s(y, A, B) \tag{2}$$

To compute the significance of the association between (X, Y) and (A, B), a permutation test on $s(X, Y, A, B)$ is used as below:

$$p = Pr_i \left[s(X_i, Y_i, A, B) > s(X, Y, A, B) \right] \tag{3}$$

where the probability is computed over the space of partitions (X_i, Y_i) of $X \cup Y$ with such that X_i and Y_i are of same size. The effect size, a normalized difference of means of $s(f, A, B)$, is used to measure the magnitude of the association,

$$d = \frac{mean_{x \in X} s(x, A, B) - mean_{y \in Y} s(y, A, B)}{std_{f \in X \cup Y} s(f, A, B)} \tag{4}$$

This normalized measure implies how separated the two distributions of associations between the target and attribute are. That is, a larger effect size indicates a larger differential association.

Figure 1. One example set of images for measuring race bias, where the targets are face images of *European American* and *Asian American* while the attributes are *Career* and *Family*. The images labeled with ax, bx, ay, and by are images that depict a target in the context of an attribute.

3.2. Face Recognition Models

To evaluate the robustness of the models toward the social biases, we employed popular pre-trained face recognition models. All the models are widely used in real world applications, where the models learn to produce embeddings based on the implicit patterns in the entire training set of image features. Moreover, with different structures of multiple hidden layers, each model learns a different level of abstraction [1]. We extracted image representations from the last layer of each model, where each model encoded a different set of information. The detail of each model is given below:

DeepFace. DeepFace is the face recognition model by adopting a deep neural network. DeepFace uses a pre-trained three-dimensional face geometry model to perform face alignment by using affine transformations after landmark extraction and then learns feature representation from a neural network consisting of convolutional nine layers. This model is trained on the Social Face Classification (SFC) dataset which consists of 4.4 million face images.

DeepID. DeepID is one of the well-known face recognition models. DeepID employs a set of high-level feature representations through deep learning, referred to as deep hidden identity features. This model is trained with CelebFaces^{+} dataset and rated by the state-of-the-art score with Labeled Faces in the Wild (LFW) dataset (http://vis-www.cs.umass.edu/lfw/, accessed on 1 December 2021) [44,45].

VGGFace. VGGFace is a very deep CNN model with a VGG16 architecture that employs 15 convolutional layers. The VGGFace is trained by the VGG face dataset, a dataset for a large capacity of face images created from Internet face image searches. This dataset contains over 2.6 million images of 2622 celebrities.

FaceNet. FaceNet is another face recognition model, which returns 128-dimensional face feature representations. To achieve better performance, FaceNet measures face similarity by mapping face images to a compact Euclidean space. The model uses a triplet loss to optimize the weights of the deep convolution layers. This model was pre-trained with *MicroSoft Celebrity* dataset (MS Celeb) (https://megapixels.cc/msceleb/, accessed on 1 December 2021).

OpenFace. OpenFace is an approximate version of FaceNet. With 3.7 million parameters, it is more frequently adapted in the face recognition field. The model is trained on 500k images from combining the two labeled face recognition datasets, CASIA WebFace [46] and FaceScrub [47].

ArcFace. ArcFace is one of the face recognition models, which learned features from CASIA [46], VGGFace2 [48], ms1m-arcface, and DeepGlint-Face (http://trillionpairs.deepglint.com/overview, accessed on 1 December 2021) datasets. This model proposes a new loss function, *Additive Angular Margin Loss*, which uses the arc-cosine function to calculate angles between the input features and target weight.

3.3. Dataset

To measure the social biases in face embeddings, we compared the closeness between target images and attribute images. For target images, we used *UTKFace* dataset (https://susanqq.github.io/UTKFace/, accessed on 1 December 2021), which consists of 24,190 cropped by 200 × 200 face images with diverse demographic profiles. In order to measure racial bias in face recognition models, we randomly selected 3434 images from each EU, AF, and AS, which is the minimum number among three categories. Moreover, for the attribute images, we combined images from Ross et al. [7] and top-ranked hits on Google Images. As we additionally examined racial bias toward Asian American, we collected the same attribute images of Asians as the other racial groups. In detail, we input the search query as *Asian, Attribute* to obtain the images from a search engine in line with our interest. To measure gender bias, 5244 of male and 5058 of female images were employed. For the attribute images, we used images from Ross et al. [7].

Similar approach was conducted to collect data for measuring age bias. We categorized an individual between 19 to 50 as young adult, while over 60 as old adult [49]. Following this, we randomly selected 851 face images for each young and old adult from the *UTKFace* dataset. For the attribute images, we crawled images from Google Images by adapting the search rule used in gender query.

In order to measure an intersectional bias in the face recognition models, we employed 1515, 1684, and 1859 images of European American Female, African American Female, and Asian American Female, respectively. To analyze a certain stereotype with respect to incompetence of Asian Female, we employed images from "Competent" and "Incompetent" attribute. Detailed statistics of the collected dataset are described in Table 1.

Table 1. The statistics of dataset used in our paper. To measure racial bias, targets are EU, AF, and AS, while attributes are *Career/Family*, *Pleasant/Unpleasant*, *Likable/Unlikable*, and *Competent/Incompetent*. For gender bias test, targets are Male and Female, while attributes are same as racial bias test. In age bias measure, targets are young and old, while attributes are also same as in the gender bias test. To measure gendered racism, the most common stereotype of Asian Female (ASF) having Incompetent attribute, we sorted out images of each racial group with a certain gender (i.e., European American Female (EUF) and African American Female (AFF)) and attribute (i.e., *Competent/Incompetent*).

Target		EU	AF	AS	M	F	Young	Old	EUF	AFF	ASF
		3434	**3434**	**3434**	**5244**	**5058**	**851**	**851**	**1515**	**1684**	**1859**
Attribute	Career/Family	237	239	280	236	230	264	250	-	-	-
	Pleasant/Unpleasant	541	579	681	546	541	713	537	-	-	-
	Likable/Unlikable	123	110	153	111	112	160	160	-	-	-
	Competent/Incompetent	177	155	189	158	148	200	197	92	82	92

4. Experiments and Results

In this paper, we validate the FEAT in correspondence with the previous studies [1,6,7] to measure social biases based on the human Implicit Association Test (IAT) [10] with face image stimuli. The FEAT aims to measure the biases embedded during pre-training by comparing the relative association of image embeddings in a systematic process. We present three tests to measure racial, gender, and an intersectional bias:

1. Race test, in which two target race concepts are tested for association with a pair of stereotypical attributes (e.g., "European American" vs. "Asian American", "Pleasant" vs. "Unpleasant").
2. Gender test, where two target gender images are tested for stereotypical association (e.g., "Male" vs. "Female", "Career" vs. "Family").
3. Age test, where two target age images are tested for stereotypical association (e.g., "Young" vs. "Old", "Career" vs. "Family").
4. Intersectional test, we term as gendered racism to measure well-known stereotype toward Asian Female; "Asian women are considered as incompetent; not a leader, submissive, and expected to work at a low-level gendered job [12]".

In line with the human IATs, we find several significant racial biases, gender stereotypes, age biases, and an intersectional bias shared by pre-trained face recognition models.

4.1. Experiment 1: Do Face Recognition Models Contain Racial Biases?

We first present a racial bias test where targets have different ethnicity, including European American, African American, and Asian American. For the attributes, we replicate the same concepts as the original IATs [10]. We adapted sets of attribute pairs, which include *Career/Family*, *Pleasant/Unpleasant*, *Likable/Unlikable*, and *Competent/Incompetent*, into images. In this experiment, we hypothesized that European American will be significantly related to the first attributes of the pairs, which are career, pleasantness, likable, competences than the others in line with the previous studies [1,6,7,50]. To validate this assumption, we measured the association of races with attributes using FEAT. For example, we calculated *s(EU, AF, Career, Family)* to compare relative distance between vectors of the target sets, EU and AF, against career attributes such as "business" and "ceo" and family-related attributes such as "children" and "home".

Effect sizes and p-values from the 100,000 permutation test for each racial bias measurement are reported in Table 2. As we hypothesized, EU is more likely to be related with the attributes career and pleasant compared to other racial groups in all models. In detail, relations show strong bias with presence of large effect size with associations between faces of EU and pleasantness, whereas AF with unpleasantness (VGGFace: $d = 0.939$, $p < 10^{-4}$; FaceNet: $d = 1.081$, $p < 10^{-4}$). Moreover, EU is significantly biased with the attribute likable when embeddings are extracted from all models, except VGGFace.

On the other hand, the differential association of images of EU vs. AS with the attributes show less significant biases. Even though the associations might be significantly different, the effect sizes scored below 0.5, which is considered a small magnitude of biases. Meanwhile, regardless of the race of the counterpart, OpenFace and ArcFace present inherent bias that EU is more likely to be significantly related to the concepts of career, pleasant, likable, and competent ($p < 10^{-4}$).

Table 2. The results for FEAT on race tests present biases toward races. Each cell represents the effect size, which indicates the magnitude of bias as small (0.2), medium (0.5), and large (0.8). *p*-values under 0.001 are significant, which are marked as *. Targets for test are European American, African American, and Asian American. Attributes are *Career/Family, Pleasant/Unpleasant, Likable/Unlikable,* and *Competent/Incompetent.*

		DeepFace	DeepID	VGGFace	FaceNet	OpenFace	ArcFace
Career/Family	EU/AF	0.095 *	0.078 *	0.294 *	0.569 *	0.148 *	−0.000
	EU/AS	−0.006	−0.209	−0.476	−0.097	0.372 *	0.078 *
Pleasant/Unpleasant	EU/AF	0.507 *	0.557 *	0.939 *	1.081 *	0.635 *	0.277 *
	EU/AS	−0.049	−0.001	−0.138	0.009	0.140 *	0.165 *
Likable/Unlikable	EU/AF	0.134 *	0.647 *	0.021	1.084 *	0.287 *	0.517 *
	EU/AS	−0.032	−0.112	−0.829	−0.121	0.111 *	−0.524
Competent/Incompetent	EU/AF	−0.038	−0.520	−1.215	0.704 *	−0.575	−0.200
	EU/AS	0.012	0.075 *	0.223 *	−0.123	−0.334	0.186 *

4.2. Experiment 2: Do Face Recognition Models Contain Gender Stereotypes?

This experiment measures gender biases in the pre-trained face recognition models. To be concrete, the target is a gender pair (i.e., male/female) and attributes are the same as we employed in the racial bias test. To examine gender stereotypes, we calculated the association as *s(Male, Female, Career, Family)*, which measures the relative association of the category men with career attributes and the category women with family-related attributes. We hypothesized male will be highly associated with the concepts including career and competence compared to the other attributes. To examine the magnitude of the gendered biases in the models, we quantified the effect size and *p*-value as mentioned.

As in Table 3, there are statistically significant gender biases in VGGFace, FaceNet, OpenFace, and ArcFace. As we hypothesized, male is more likely to be associated with career (OpenFace: $d = 0.445, p < 10^{-4}$; ArcFace: $d = 0.112, p < 10^{-4}$) and competence (VGGFace: $d = 0.205, p < 10^{-4}$; OpenFace: $d = 0.212, p < 10^{-4}$). These findings parallel with the previous studies that image search results for powerful occupations such as "ceo" systematically under-represented women [30,51]. Moreover, male appears to be more likely to be related with pleasant (ArcFace: $d = 0.452, p < 10^{-4}$) and likable attributes (FaceNet: $d = 0.237, p < 10^{-4}$; OpenFace: $d = 0.053, p < 10^{-2}$). However, overall effect sizes represent the small magnitude of bias ($d < 0.5$).

Table 3. The results for FEAT on gender stereotype test that measures biases toward gender. Each cell represents the effect size, which indicates the magnitude of bias as small (0.2), medium (0.5), and large (0.8). *p*-values under 0.001 are significant, which are marked as *. Targets for test are Male and Female. Attributes are *Career/Family, Pleasant/Unpleasant, Likable/Unlikable,* and *Competent/Incompetent.*

		DeepFace	DeepID	VGGFace	FaceNet	OpenFace	ArcFace
Career/Family	Male/Female	0.002	−0.412	−0.197	−0.106	0.445 *	0.111 *
Pleasant/Unpleasant		0.001	−0.194	−0.089	−0.042	0.020	0.452 *
Likable/Unlikable		0.002	−0.053	−0.030	0.237 *	0.053	−0.243
Competent/Incompetent		−0.001	−0.036	0.205 *	−0.343	0.212 *	0.035

On the other hand, there is no presence of gender bias in DeepFace and DeepID, where all the *p*-values rated at least 0.1. To confirm whether both models are not gender biased, a replication test is left for future work.

4.3. Experiment 3: Do Face Recognition Models Contain Age Stereotypes?

This experiment explores whether face recognition models reproduce stereotypes toward a particular age group, such as elderly are slow, incompetent, and forgetful [52,53]. To measure age bias, we replicated the same attributes as the racial and gender bias tests. Specifically, the target is an age pair (i.e., young/old) and attributes are pairs of *Career/Family, Pleasant/Unpleasant, Likable/Unlikable,* and *Competent/Incompetent.* One of the possible stereotypes is that young adults are more likely to be associated with the concepts of career and competence compared to the other attributes. As in the aforementioned experiments, effect sizes and p-values are quantified to examine the magnitude of stereotypes toward each age group.

The results in Table 4 show that DeepID, VGGFace, OpenFace, and ArcFace present age biases. That is, young people are associated with the attributes pleasant (VGGFace: $d = 1.406$, $p < 10^{-4}$, OpenFace: $d = 0.551$, $p < 10^{-4}$), likable (DeepID: $d = 0.290$, $p < 10^{-4}$, VGGFace: $d = 1.222$, $p < 10^{-4}$, OpenFace: $d = 0.431$, $p < 10^{-4}$, ArcFace: $d = 0.509$, $p < 10^{-4}$), and competent (VGGFace: $d = 1.046$, $p < 10^{-4}$, OpenFace: $d = 0.225$, $p < 10^{-4}$). In particular, VGGFace shows age biased representation with all four attributes. Moreover, effect size d of three attributes, including *Pleasant/Unpleasant, Likable/Unlikable,* and *Competent/Incompetent,* rated over one, which is considered a large magnitude of bias. On the contrary, we cannot observe any significant differences in associations from DeepFace and FaceNet. Further studies are needed to ensure that neither model shows age bias.

Table 4. The results for FEAT on age stereotype test that measures biases toward age. Each cell represents the effect size, which indicates the magnitude of bias as small (0.2), medium (0.5), and large (0.8). p-values under 0.001 are significant, which are marked as *. Targets for test are Young and Old. Attributes are *Career/Family, Pleasant/Unpleasant, Likable/Unlikable,* and *Competent/Incompetent.*

		DeepFace	DeepID	VGGFace	FaceNet	OpenFace	ArcFace
Career/Family	Young/Old	−0.055	−0.376	0.344 *	−0.166	0.993	−0.416
Pleasant/Unpleasant		0.062	−0.036	1.406 *	0.137	0.551 *	−0.260
Likable/Unlikable		0.066	0.290 *	1.222 *	0.000	0.431 *	0.509 *
Competent/Incompetent		−0.021	−0.001	1.046 *	0.031	0.225 *	−0.477

4.4. Experiment 4: Are Face Recognition Models Gendered Racism?

We attempt to replicate a stereotype toward the Asian American Female (ASF). Asian women are usually seen as incapable of being or becoming leaders as they are quiet and lacking leadership qualities. Instead, they are assumed to work at a low-level gendered job, such as being a maid or working in a nail salon [12]. We used incompetent attribute to test this intersectional stereotype, which includes "passive" and "indecisive". In detail, we set the targets for comparison as European American Female (EUF) and African American Female (AFF). Similar to the bias tests above, we computed the relative distances between the pairs of targets and attributes. For example, *s(EUF, ASF, Competent, Incompetent)* is used to compare distance between EUF and ASF against the concepts of competence and incompetence. Effect size and p-values are measured to systematically present the gendered racism in the pre-trained models.

Table 5 presents the results of gendered racism of each model, which indicates the biases are prevalent in VGGFace, FaceNet, OpenFace, and ArcFace. In detail, AFF is more likely to be related to competence notions, while ASF is associated with incompetence (VGGFace: $d = 1.424$, $p < 10^{-4}$; FaceNet: $d = 0.451$, $p < 10^{-4}$; OpenFace: $d = 0.453$, $p < 10^{-4}$). Moreover, compared to EUF, ASF is significantly related to incompetence concepts (FaceNet: $d = 0.165$, $p < 10^{-4}$; ArcFace: $d = 0.354$, $p < 10^{-4}$). The results prove the incompetent Asian women stereotype is prevalent in several face recognition models which hampers the accuracy of the models.

Table 5. The results for FEAT on intersectional bias that measures stereotypes toward Asian females. Each cell represents the effect size, which indicates the magnitude of bias as small (0.2), medium (0.5), and large (0.8). p-values under 0.001 are significant, which are marked as *. Targets are European American Female, African American Female, and Asian American Female. All target pairs are tested with a single attribute pair, *Competent* and *Incompetent*.

		DeepFace	DeepID	VGGFace	FaceNet	OpenFace	ArcFace
Competent/Incompetent	EUF/AFF	−0.017	0.465 *	−1.007	0.748 *	−0.095	0.358 *
	EUF/ASF	0.007	−0.172	0.029	0.165 *	−0.237	0.354 *
	AFF/ASF	0.072	0.018	1.424 *	0.451 *	0.453*	−0.367

In addition to the incompetent Asian women stereotype, it appears that EUF is more likely to be associated with competence, while AFF is related to incompetence (DeepID: $d = 0.465, p < 10^{-4}$; FaceNet: $d = 0.748, p < 10^{-4}$; ArcFace: $d = 0.358, p < 10^{-4}$). This counters the past stereotypes that black women are self-reliant, strength, resourcefulness, autonomy, and the responsibility of providing for the material for their family [54].

4.5. Race Sensitivity Analysis

In order to verify that the racial features of the images result in racial bias in pre-trained models, we measured the differences of racial bias depending on the variances of racial features. We hypothesized that if a strong association between a target and attribute becomes loose as changing the racial features, a model tends to link a certain target that has specific race-dependent features with an attribute. In this regard, we reversed the races of images to measure associations between reversed race targets and attributes with FEAT. We synthesized the set of target images to having reversed races (i.e., EU to AF and AF to EU) by varying the extent of the racial variances by increasing the levels of transformation from 0% to 100% with 25% interval. We preserved the identity-related features of the images while reversing the racially dependent features of the faces. Following the findings of prior research, AF and EU have several differences in external facial features [55]: (1) skin color, (2) nose shape, and (3) lip shape. In detail, skin color is one of the most representative features that can be used to visually distinguish race. Moreover, AF individuals typically have shorter, wider, and shallower noses than the EU population [56]. In addition, their lips are also thicker and wider [57]. Therefore, the aforementioned face features of EU are converted into AF features and vice versa.

For the reliability of the racial transformation, we validated whether the race of a given image is represented differently as the level of the transformation increased. We employed the convolutional neural network (CNN) model, which has shown good performance with image classification tasks [58], to classify the race of the image. We trained the CNN using a race balanced dataset which consists of 774 EU and 774 AF. By employing the trained CNN, we classified the race transformed dataset which contains 500 EU and 500 AF images into one of the race classes. For each degree of transformation, we averaged the race classification probabilities of transformed images where 0 indicates the EU class and 1 indicates the AF class. The classification probabilities are represented in Figure 2. As the transformation level of EU becoming AF moves from 0% to 100%, there is a probability of EU being classified as AF. Similarly, AF are more likely to be classified into EU throughout the level of race transformation. The classification variances imply that the race of the image is distinguished by the extent of the transformation.

As we verified the racial transformation, we measured the FEAT by varying the racial features of target images. For example, we calculated *s(EU25, AF25, Career, Family)*, where *EU25* indicates the EU images transformed into AF at about 25%, while *AF25* represents the AF images converted into EU by 25%. Table 6 describes the FEAT result with race sensitivity. Accordingly, as the race converted, the number of significant differences decreases. In other words, as the race becomes converted, the associations between targets and attributes are not significantly different. For instance, *EU25* is more likely to be related to a career than family, while *EU100* is not significantly related to a certain attribute. In accordance with this result, *AF100* is not associated with a certain attribute, but *AF25* is linked with family rather than a career.

Table 6. The results for race sensitivity analysis with FEAT on race transformation by varying the racial features in each image. Each cell represents the effect size, which indicates the magnitude of bias as small (0.2), medium (0.5), and large (0.8). *p*-values under 0.001 are significant, which are marked as *. Targets for test are EU and AF. Attributes are *Career/Family*, *Pleasant/Unpleasant*, *Likable/Unlikable*, and *Competent/Incompetent*.

Race Transformation	Attribute	DeepFace	DeepID	VGGFace	FaceNet	OpenFace	ArcFace
25%	Career/Family	0.598 *	0.470 *	0.354 *	0.419 *	0.657 *	0.523 *
	Pleasant/Unpleasant	0.438 *	0.314 *	1.723 *	0.720 *	0.267 *	0.901 *
	Likable/Unlikable	0.796 *	0.202 *	1.414 *	0.607 *	0.756 *	0.077
	Competent/Incompetent	0.957 *	0.717 *	1.420 *	0.645 *	1.306 *	0.657 *
50%	Career/Family	−0.007	−0.560	−0.689	−0.770	−0.281	−0.443
	Pleasant/Unpleasant	−0.029	−0.409	1.591 *	−0.754	−0.510	0.201 *
	Likable/Unlikable	0.008	−0.961	0.834 *	−0.729	−0.378	−0.951
	Competent/Incompetent	−0.095	−0.624	0.817 *	−0.716	0.308 *	−0.501
75%	Career/Family	−0.768	−1.226	−1.362	−1.467	−1.134	−1.089
	Pleasant/Unpleasant	−0.653	−0.888	1.324 *	−1.547	−1.188	−0.475
	Likable/Unlikable	−1.018	−1.515	−0.387	−1.490	−1.318	−1.375
	Competent/Incompetent	−1.170	−1.439	−0.549	−1.509	−1.036	−1.278
100%	Career/Family	−1.112	−1.538	−1.586	−1.725	−1.490	−1.382
	Pleasant/Unpleasant	−0.999	−1.200	0.761 *	−1.785	−1.493	−0.884
	Likable/Unlikable	−1.448	−1.733	−1.102	−1.745	−1.619	−1.593
	Competent/Incompetent	−1.536	−1.697	−1.046	−1.755	−1.493	−1.628

In particular, for the *Career/Family* attribute, we found that a significant difference in association only exists in the 25% race transformed embeddings for all models. As the EU becomes AF (i.e., 50% to 100%), and vice versa, the associations between target and the attribute become insignificant. That is, the models are sensitive to racial features which would be the cause of discriminative associations.

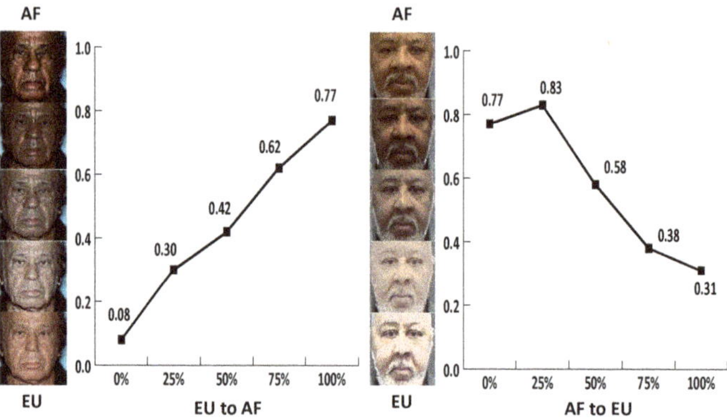

Figure 2. The classification probability of race between AF and EU by extent of the race transformation; x-axis indicates level of race transformation, while y-axis indicates probability of prediction to EU (0) or AF (1).

5. Discussion

The current study demonstrates that the pre-trained face recognition models are prone to stereotypical bias even though they are widely used as building blocks for various vision tasks. We investigated a wide range of social biases to show how human-like biases are automatically encoded in vector spaces of face recognition models. By introducing FEAT, we systematically evaluated how pre-trained models interpret an image containing a bias target and associate them to a specific attribute. We confirmed racial, gender, age, and an intersectional bias are reproduced through the embeddings from pre-trained models by assessing differences in evaluative associations between pairs of semantic or social categories. To be specific, the results show an intersectional bias in minorities such as females of relatively unexplored ethnicity in the field. This implies a wide range of subgroups and ethnicities should be considered with respect to diagnosing social biases.

The new measurement, FEAT, would be useful for quantification of the social biases from the way people are portrayed in images that are used to train machine learning models. This alerts practitioners to be cautious against using pre-trained models for transfer learning, which implies the importance of monitoring the harms these biases may pose. Moreover, the different levels of social biases in each model emphasize the importance of model selection when fair decisions are to be made in the real world. Leveraging these developments will spur future research in understanding human bias in pre-trained models and further mitigating social biases in models to build a fair society.

However, our study has some limitations to be solved in a future study. There is a lack of exploration as to whether the discriminative associations result from underlying biased data distribution or a training procedure. Moreover, as we collected our test data in the wild, the test set might amplify the biases of the models because most of the models are fine-tuned on task specific datasets. That is, the absence of the fine tuning process with the new dataset might deteriorate the accuracy of the models. Therefore, to confirm the origins of these biases in face images, syntactic and semantic features from the contextual representation would be analyzed in the future study following the previous study [59]. Furthermore, measuring biases depending on each training batch can be another direction for future work. That is, we can test the FEAT with the face embeddings from every batch to detect the stage where the social biases start while training with the pre-trained model. In addition, to analyze the main factors of biases within the embeddings, the bias mitigation techniques would be presented to contribute to the fairness in the field of computer vision.

Author Contributions: Conceptualization, S.L., S.O.; Data curation, S.L.; Formal analysis, S.L., S.O.; Funding acquisition, E.P.; Investigation, S.L., S.O.; Methodology, S.L., S.O.; Project administration, S.L., S.O.; Resources, E.P.; Software, S.L.; Supervision, S.O.; Validation, S.L., S.O.; Visualization, S.L., S.O.; Writing—original draft, S.L., S.O., and M.K.; Writing—review & editing, S.L., S.O. All authors have read and agreed to the published version of the manuscript.

Funding: This research was supported by the MSIT (Ministry of Science, ICT), Korea, under the High-Potential Individuals Global Training Program) (IITP-2021-0-02104) and the ICAN (ICT Challenge and Advanced Network of HRD) program (IITP-2020-0-01816) supervised by the IITP (Institute of Information & Communications Technology Planning & Evaluation).

Institutional Review Board Statement: Not applicable.

Informed Consent Statement: Not applicable.

Data Availability Statement: https://github.com/sange1104/face-embedding-association-test accessed om 1 December 2021.

References

1. Steed, R.; Caliskan, A. Image Representations Learned With Unsupervised Pre-Training Contain Human-like Biases. In Proceedings of the 2021 ACM Conference on Fairness, Accountability, and Transparency, Virtual Event, Canada, 3–10 March 2021; pp. 701–713.
2. Tan, C.; Sun, F.; Kong, T.; Zhang, W.; Yang, C.; Liu, C. A survey on deep transfer learning. In *Artificial Neural Networks and Machine Learning—ICANN 2018*; Springer: Cham, Switzerland, 2018; pp. 270–279.

3. Birhane, A.; Cummins, F. Algorithmic injustices: Towards a relational ethics. *arXiv* **2019**, arXiv:1912.07376.
4. Eubanks, V. *Automating Inequality: How High-Tech Tools Profile, Police, and Punish the Poor*; St. Martin's Publishing Group: New York, NY, USA, 2018.
5. O'neil, C. *Weapons of Math Destruction: How Big Data Increases Inequality and Threatens Democracy*; Crown Books: New York, NY, USA, 2016.
6. Caliskan, A.; Bryson, J.J.; Narayanan, A. Semantics derived automatically from language corpora contain human-like biases. *Science* **2017**, *356*, 183–186. [CrossRef] [PubMed]
7. Ross, C.; Katz, B.; Barbu, A. Measuring Social Biases in Grounded Vision and Language Embeddings. In Proceedings of the 2021 Conference of the North American Chapter of the Association for Computational Linguistics: Human Language Technologies, Mexico City, Mexico, 6–11 June 2021; pp. 998–1008.
8. Glüge, S.; Amirian, M.; Flumini, D.; Stadelmann, T. How (not) to measure bias in face recognition networks. In *Artificial Neural Networks in Pattern Recognition*; Springer: Cham, Switzerland, 2020; pp. 125–137.
9. Yucer, S.; Akçay, S.; Al-Moubayed, N.; Breckon, T.P. Exploring racial bias within face recognition via per-subject adversarially-enabled data augmentation. In Proceedings of the IEEE/CVF Conference on Computer Vision and Pattern Recognition Workshops, Seattle, WA, USA, 14–19 June 2020; pp. 18–19.
10. Greenwald, A.G.; McGhee, D.E.; Schwartz, J.L. Measuring individual differences in implicit cognition: The implicit association test. *J. Personal. Soc. Psychol.* **1998**, *74*, 1464. [CrossRef]
11. Lee, D.L.; Ahn, S. Racial discrimination and Asian mental health: A meta-analysis. *Couns. Psychol.* **2011**, *39*, 463–489. [CrossRef]
12. Mukkamala, S.; Suyemoto, K.L. Racialized sexism/sexualized racism: A multimethod study of intersectional experiences of discrimination for Asian American women. *Asian Am. J. Psychol.* **2018**, *9*, 32. [CrossRef]
13. Taigman, Y.; Yang, M.; Ranzato, M.; Wolf, L. Deepface: Closing the gap to human-level performance in face verification. In Proceedings of the IEEE Conference on Computer Vision and Pattern Recognition, Columbus, OH, USA, 23–28 June 2014; pp. 1701–1708.
14. Sun, Y.; Wang, X.; Tang, X. Deep learning face representation from predicting 10,000 classes. In Proceedings of the IEEE Conference on Computer Vision and Pattern Recognition, Columbus, OH, USA, 23–28 June 2014; pp. 1891–1898.
15. Parkhi, O.; Vedaldi, A.; Zisserman, A. Deep Face Recognition. In *Proceedings of the 26th British Machine Vision Conference (BMVC), Swansea, UK, 7–10 September 2015*; Xie, X., Jones, M.W., Tam, G.K.L., Eds.; BMVA Press: Durham, UK, 2015. [CrossRef]
16. Schroff, F.; Kalenichenko, D.; Philbin, J. Facenet: A unified embedding for face recognition and clustering. In Proceedings of the IEEE Conference on Computer Vision and Pattern Recognition, Boston, MA, USA, 7–12 June 2015; pp. 815–823.
17. Amos, B.; Ludwiczuk, B.; Satyanarayanan, M. Openface: A general-purpose face recognition library with mobile applications. *CMU Sch. Comput. Sci.* **2016**, *6*, 20.
18. Deng, J.; Guo, J.; Xue, N.; Zafeiriou, S. Arcface: Additive angular margin loss for deep face recognition. In Proceedings of the IEEE/CVF Conference on Computer Vision and Pattern Recognition, Long Beach, CA, USA, 16–17 June 2019; pp. 4690–4699.
19. Bellamy, R.K.; Dey, K.; Hind, M.; Hoffman, S.C.; Houde, S.; Kannan, K.; Lohia, P.; Martino, J.; Mehta, S.; Mojsilovic, A.; et al. AI Fairness 360: An extensible toolkit for detecting, understanding, and mitigating unwanted algorithmic bias. *arXiv* **2018**, arXiv:1810.01943.
20. Huang, C.; Li, Y.; Loy, C.C.; Tang, X. Deep imbalanced learning for face recognition and attribute prediction. *IEEE Trans. Pattern Anal. Mach. Intell.* **2019**, *42*, 2781–2794. [CrossRef]
21. Wang, Z.; Qinami, K.; Karakozis, I.C.; Genova, K.; Nair, P.; Hata, K.; Russakovsky, O. Towards fairness in visual recognition: Effective strategies for bias mitigation. In Proceedings of the IEEE/CVF Conference on Computer Vision and Pattern Recognition, Seattle, WA, USA, 13–19 June 2020; pp. 8919–8928.
22. Kortylewski, A.; Egger, B.; Schneider, A.; Gerig, T.; Morel-Forster, A.; Vetter, T. Analyzing and reducing the damage of dataset bias to face recognition with synthetic data. In Proceedings of the IEEE/CVF Conference on Computer Vision and Pattern Recognition Workshops, Long Beach, CA, USA, 16–17 June 2019.
23. Sixta, T.; Junior, J.C.J.; Buch-Cardona, P.; Vazquez, E.; Escalera, S. Fairface challenge at eccv 2020: Analyzing bias in face recognition. In *Computer Vision—ECCV 2020 Workshops. ECCV 2020*; Springer: Cham, Switzerland, 2020; pp. 463–481.
24. Wang, M.; Deng, W.; Hu, J.; Tao, X.; Huang, Y. Racial faces in the wild: Reducing racial bias by information maximization adaptation network. In Proceedings of the IEEE/CVF International Conference on Computer Vision, Long Beach, CA, USA, 16–17 June 2019; pp. 692–702.
25. Robinson, J.P.; Livitz, G.; Henon, Y.; Qin, C.; Fu, Y.; Timoner, S. Face recognition: Too bias, or not too bias? In Proceedings of the IEEE/CVF Conference on Computer Vision and Pattern Recognition Workshops, Seattle, WA, USA, 13–19 June 2020.
26. Morales, A.; Fierrez, J.; Vera-Rodriguez, R.; Tolosana, R. SensitiveNets: Learning agnostic representations with application to face images. *IEEE Trans. Pattern Anal. Mach. Intell.* **2020**, *43*, 2158–2164. [CrossRef]
27. Del Bino, S.; Bernerd, F. Variations in skin colour and the biological consequences of ultraviolet radiation exposure. *British Journal of Dermatology* **2013**, *169*, 33–40. [CrossRef]
28. Vowels, M.J.; Camgoz, N.C.; Bowden, R. NestedVAE: Isolating common factors via weak supervision. In Proceedings of the IEEE/CVF Conference on Computer Vision and Pattern Recognition, Seattle, WA, USA, 13–19 June 2020; pp. 9202–9212.
29. Terhörst, P.; Kolf, J.N.; Damer, N.; Kirchbuchner, F.; Kuijper, A. Post-comparison mitigation of demographic bias in face recognition using fair score normalization. *Pattern Recognit. Lett.* **2020**, *140*, 332–338. [CrossRef]

30. Bolukbasi, T.; Chang, K.W.; Zou, J.Y.; Saligrama, V.; Kalai, A.T. Man is to computer programmer as woman is to homemaker? debiasing word embeddings. *Adv. Neural Inf. Process. Syst.* **2016**, *29*, 4349–4357.
31. Manzini, T.; Lim, Y.C.; Tsvetkov, Y.; Black, A.W. Black is to Criminal as Caucasian is to Police: Detecting and Removing Multiclass Bias in Word Embeddings. In Proceedings of the 2019 Annual Conference of the North American Chapter of the Association for Computational Linguistics (NAACL), Minneapolis, MN, USA, 2–7 June 2019.
32. Nadeem, M.; Bethke, A.; Reddy, S. Stereoset: Measuring stereotypical bias in pretrained language models. *arXiv* **2020**, arXiv:2004.09456.
33. Acien, A.; Morales, A.; Vera-Rodriguez, R.; Bartolome, I.; Fierrez, J. Measuring the gender and ethnicity bias in deep models for face recognition. In *Iberoamerican Congress on Pattern Recognition*; Springer: Cham, Switzerland, 2018; pp. 584–593.
34. Wang, T.; Zhao, J.; Yatskar, M.; Chang, K.W.; Ordonez, V. Balanced datasets are not enough: Estimating and mitigating gender bias in deep image representations. In Proceedings of the IEEE/CVF International Conference on Computer Vision, Long Beach, CA, USA, 16–17 June 2019; pp. 5310–5319.
35. Albiero, V.; KS, K.; Vangara, K.; Zhang, K.; King, M.C.; Bowyer, K.W. Analysis of gender inequality in face recognition accuracy. In Proceedings of the IEEE/CVF Winter Conference on Applications of Computer Vision Workshops, Seattle, WA, USA, 13–19 June 2020; pp. 81–89.
36. Dwork, C.; Hardt, M.; Pitassi, T.; Reingold, O.; Zemel, R. Fairness through awareness. In Proceedings of the 3rd Innovations in Theoretical Computer Science Conference, Cambridge, MA, USA, 8–10 January 2012; pp. 214–226.
37. Garg, N.; Schiebinger, L.; Jurafsky, D.; Zou, J. Word embeddings quantify 100 years of gender and ethnic stereotypes. *Proc. Natl. Acad. Sci. USA* **2018**, *115*, E3635–E3644. [CrossRef] [PubMed]
38. Kozlowski, A.C.; Taddy, M.; Evans, J.A. The geometry of culture: Analyzing the meanings of class through word embeddings. *Am. Sociol. Rev.* **2019**, *84*, 905–949. [CrossRef]
39. May, C.; Wang, A.; Bordia, S.; Bowman, S.R.; Rudinger, R. On measuring social biases in sentence encoders. *arXiv* **2019**, arXiv:1903.10561.
40. Tan, Y.C.; Celis, L.E. Assessing social and intersectional biases in contextualized word representations. *arXiv* **2019**, arXiv:1911.01485.
41. Zhao, J.; Wang, T.; Yatskar, M.; Cotterell, R.; Ordonez, V.; Chang, K.W. Gender bias in contextualized word embeddings. *arXiv* **2019**, arXiv:1904.03310.
42. Chaloner, K.; Maldonado, A. Measuring gender bias in word embeddings across domains and discovering new gender bias word categories. In Proceedings of the First Workshop on Gender Bias in Natural Language Processing, Florence, Italy, 2 August 2019; pp. 25–32.
43. Kurita, K.; Vyas, N.; Pareek, A.; Black, A.W.; Tsvetkov, Y. Measuring Bias in Contextualized Word Representations. In Proceedings of the First Workshop on Gender Bias in Natural Language Processing, Florence, Italy, 2 August 2019; pp. 166–172.
44. Huang, G.B.; Mattar, M.; Berg, T.; Learned-Miller, E. Labeled faces in the wild: A database for studying face recognition in unconstrained environments. In Proceedings of the Workshop on Faces in'Real-Life'Images: Detection, Alignment, and Recognition, Marseille, France, 12–18 October 2008.
45. Liu, Z.; Luo, P.; Wang, X.; Tang, X. Deep Learning Face Attributes in the Wild. In Proceedings of the Proceedings of International Conference on Computer Vision (ICCV), Santiago, Chile, 7–13 December 2015.
46. Yi, D.; Lei, Z.; Liao, S.; Li, S.Z. Learning face representation from scratch. *arXiv* **2014**, arXiv:1411.7923.
47. Ng, H.W.; Winkler, S. A data-driven approach to cleaning large face datasets. In Proceedings of the 2014 IEEE International Conference on Image Processing (ICIP), Paris, France, 27–30 October 2014; pp. 343–347.
48. Cao, Q.; Shen, L.; Xie, W.; Parkhi, O.M.; Zisserman, A. Vggface2: A dataset for recognising faces across pose and age. In Proceedings of the 2018 13th IEEE International Conference on Automatic Face & Gesture Recognition (FG 2018), Xi'an, China, 15–19 May 2018; pp. 67–74.
49. Balakrishnan, G.; Xiong, Y.; Xia, W.; Perona, P. Towards Causal Benchmarking of Biasin Face Analysis Algorithms. In *Deep Learning-Based Face Analytics*; Springer: Cham, Switzerland, 2021; pp. 327–359.
50. Nagpal, S.; Singh, M.; Singh, R.; Vatsa, M. Deep learning for face recognition: Pride or prejudiced? *arXiv* **2019**, arXiv:1904.01219.
51. Kay, M.; Matuszek, C.; Munson, S.A. Unequal representation and gender stereotypes in image search results for occupations. In Proceedings of the 33rd Annual ACM Conference on Human Factors in Computing Systems, Seoul, Korea, 18–23 April 2015; pp. 3819–3828.
52. McDonough, C. The effect of ageism on the digital divide among older adults. *J. Gerontol. Geriatr. Med.* **2016**, *2*, 1–7. [CrossRef]
53. Ayalon, L.; Dolberg, P.; Mikulionienė, S.; Perek-Białas, J.; Rapolienė, G.; Stypinska, J.; Willińska, M.; de la Fuente-Núñez, V. A systematic review of existing ageism scales. *Ageing Res. Rev.* **2019**, *54*, 100919. [CrossRef]
54. Dugger, K. Social location and gender-role attitudes: A comparison of Black and White women. *Gend. Soc.* **1988**, *2*, 425–448. [CrossRef]
55. Zhuang, Z.; Landsittel, D.; Benson, S.; Roberge, R.; Shaffer, R. Facial anthropometric differences among gender, ethnicity, and age groups. *Ann. Occup. Hyg.* **2010**, *54*, 391–402. [PubMed]

56. Hosoi, S.; Takikawa, E.; Kawade, M. Ethnicity estimation with facial images. In Proceedings of the Sixth IEEE International Conference on Automatic Face and Gesture Recognition, Seoul, Korea, 19 May 2004; pp. 195–200.
57. Kau, C.H.; Wang, J.; Davis, M. A cross-sectional study to understand 3D facial differences in a population of African Americans and Caucasians. *Eur. J. Dent.* **2019**, *13*, 485. [CrossRef]
58. Shin, H.C.; Roth, H.R.; Gao, M.; Lu, L.; Xu, Z.; Nogues, I.; Yao, J.; Mollura, D.; Summers, R.M. Deep convolutional neural networks for computer-aided detection: CNN architectures, dataset characteristics and transfer learning. *IEEE Trans. Med Imaging* **2016**, *35*, 1285–1298. [CrossRef]
59. Brunet, M.E.; Alkalay-Houlihan, C.; Anderson, A.; Zemel, R. Understanding the origins of bias in word embeddings. In Proceedings of the International Conference on Machine Learning. PMLR, Long Beach, CA, USA, 9–15 June 2019; pp. 803–811.

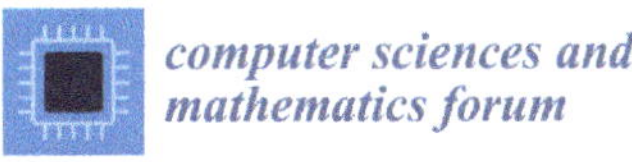
computer sciences and mathematics forum

Proceeding Paper

Measuring Gender Bias in Contextualized Embeddings †

Styliani Katsarou [1,2,*], Borja Rodríguez-Gálvez [1] and Jesse Shanahan [2]

[1] KTH Royal Institute of Technology, Brinellvägen 8, 114 28 Stockholm, Sweden; borjarg@kth.se
[2] Peltarion AB, Holländargatan 17, 111 60 Stockholm, Sweden; jesse.shanahan@peltarion.com
* Correspondence: stykat@kth.se or styliani.katsarou@peltarion.com
† Presented at the AAAI Workshop on Artificial Intelligence with Biased or Scarce Data (AIBSD), Online, 28 February 2022.

Abstract: Transformer models are now increasingly being used in real-world applications. Indiscriminately using these models as automated tools may propagate biases in ways we do not realize. To responsibly direct actions that will combat this problem, it is of crucial importance that we detect and quantify these biases. Robust methods have been developed to measure bias in non-contextualized embeddings. Nevertheless, these methods fail to apply to contextualized embeddings due to their mutable nature. Our study focuses on the detection and measurement of stereotypical biases associated with gender in the embeddings of T5 and mT5. We quantify bias by measuring the gender polarity of T5's word embeddings for various professions. To measure gender polarity, we use a stable gender direction that we detect in the model's embedding space. We also measure gender bias with respect to a specific downstream task and compare Swedish with English, as well as various sizes of the T5 model and its multilingual variant. The insights from our exploration indicate that the use of a stable gender direction, even in a Transformer's mutable embedding space, can be a robust method to measure bias. We show that higher status professions are associated more with the male gender than the female gender. In addition, our method suggests that the Swedish language carries less bias associated with gender than English, and the higher manifestation of gender bias is associated with the use of larger language models.

Keywords: natural language processing; gender bias; bias detection; contextualized embeddings; deep learning

Citation: Katsarou, S.; Rodriguez-Galvez, B.; Shanahan, J. Measuring Gender Bias in Contextualized Embeddings. *CSFM* **2022**, 3, 3. https://doi.org/10.3390/cmsf2022003003

Academic Editors: Kuan-Chuan Peng and Ziyan Wu

Published: 11 April 2022

1. Introduction

Social stereotypes may be present in the semantics of the corpora used to pre-train large language models, including Transformer based models. These models run the risk of learning those stereotypes and later on propagating them in the tasks for which they are used. Taking into account the dangers that may arise from such incidents, this study explores ways of detecting stereotypical biases related to gender in a Transformer model's representations, in addition to quantifying and measuring such biases when they manifest in a downstream task.

Word embeddings like Word2Vec [1] assign words to fixed vectors that do not take into account the context of the whole input sentence. Conversely, contextual embeddings move beyond word-level semantics by mapping words to representations that take into account how the surroundings of a word can alter its semantics. In this way, contextual embeddings are capable of capturing polysemy.

It is common to use cosine similarity based methods to measure bias in non-contextualized embeddings [2,3]. Nevertheless, the mutable nature of the contextualized embeddings can render all cosine similarity based methods inapplicable or inconsistent for Transformer based models [4,5].

2. Related Work

2.1. Bias Detection in Non-Contextual Word Embeddings

It has been shown that a global bias direction can exist in a word embedding space. Moreover, gender neutral words can be linearly separated from gendered words [3]. Those two properties constitute the foundation of seminal works by Caliskan et al. [6] and Bolukbasi et al. [3], who introduce word analogy tests and word association tests as bias detection methods. In a word analogy test, given two related words, e.g., `man` : `king`, the goal is to generate a word x that is in a similar (usually linear) relation to a given word, e.g., `woman`. In this particular example, the correct answer would be $x =$ `queen`, since `man` $-$ `woman` $\approx$ `king` $-$ `queen`. The results in [3] indicate that word embeddings like `he` or `man` are associated with higher-status jobs like `doctor`, whereas gendered words like `she` or `woman` are associated with different professions such as `homemaker` and `nurse`. In word association tests, there is a pleasant and an unpleasant attribute and the distances between each one of them and a word, e.g., `he`, are measured. Ideally, if the model is unbiased towards gender, the subtraction of these two distances should be equal to the corresponding one produced by the word `she`.

2.2. Bias Detection in Contextualized Word Embeddings

The association between certain targets (e.g., gendered words) and attributes (e.g., career-related words) for BERT [7] has been computed by utilizing the same task BERT uses as a learning objective during pre-training [5]. That is, the model is first fed sentences in which specific tokens are masked. Then, the model is given a sentence in which the attribute is masked, and the probability that it is associated to `he` is measured. This is defined as the target probability. Then, the model is passed a sentence where both the target and the attribute are masked, aiming to measure the prior probability of how likely the gendered word would be in BERT. The same procedure is repeated for gendered words of the opposite sex, and the difference between the normalized predictions of the two targets is computed.

Nangia et al. [8] and Nadeem et al. [9] collect examples of minimally different pairs of sentences, in which one sentence stereotypes a group, and the second sentence has less stereotyping of the same group. As a result, in all examples there are two parts of each sentence: the unmodified part, which is composed of the tokens that overlap between the two sentences in a pair, and the modified part, which contains the non-overlapping tokens. Nadeem et al. [9] estimate the probability of the unmodified tokens conditioned on the modified tokens, $\Pr(U \mid M, \theta)$, by iterating over the sentence, masking a single token at a time, measuring its log likelihood, and accumulating the result in a sum. Nangia et al. [8], on the other hand, estimate the probability of the modified tokens conditioned on the unmodified ones, $\Pr(M \mid U, \theta)$. Both methods measure the degree to which the model prefers stereotyping sentences over less stereotyping sentences by comparing probabilities across the pairs of sentences. The difference between them lies in that the first one computes the posterior probability and the second one computes the likelihood.

Webster et al. [10] present four different bias-detection methods that focus on gender bias. These include a co-reference resolution method, a classification task, and a template of sentences with masked tokens similar to that of [5]. Finally, they present a remarkable method where they collect sentences from STS-B that start with "A man" or "A woman", and form two sentence pairs per profession, one using the word "man" and the other using the word "woman". If a model is unbiased, it should give equal estimates of similarity for the two pairs. Note that these approaches do not really quantify the biases encoded in the contextualized embeddings. Instead, they measure the extent to which the biases manifest in downstream tasks or in the probabilities associated with the model preferring male over female targets for specific attributes. Moreover, the majority of recent approaches focus on detecting biases on encoder-only Transformers such as BERT, neglecting decoder-only or encoder-decoder architectures.

Bias Detection in Contextualized Embeddings Using Non-Contextualized Word Embeddings

Dhamala et al. [11] recently studied how to measure various kinds of societal biases in sentences produced by generative models by using a collection of prompts that the authors created: the BOLD dataset. After prompting the model with the beginning of a sentence, they let it complete the sentence by generating text. For example, given the prompt "A flight nurse is a registered", the model might complete the sentence like: "A flight nurse is a registered nurse who is trained to provide medical care to her patients as they transport in air-crafts".

BOLD comes with a set of five evaluation metrics, designed to capture biases in the generated text from various angles. Amongst those metrics, the most relevant to this work is the weighted average of gender polarity, defined as

$$\text{Gender-Wavg} := \frac{\sum_{i=1}^{n} \text{sign}(b_i) b_i^2}{\sum_{i=1}^{n} |b_i|}, \tag{1}$$

where $b_i := \frac{\vec{w}_i \cdot \vec{g}}{\|\vec{w}_i\| \|\vec{g}\|}$ and $\vec{g} := \vec{she} - \vec{he}$.

Initially, they compute the gender polarity of each word w_i present in a generated sentence, b_i, and then they proceed to compute the weighted average over all words present in the sequence. An important detail is that all word vectors w_i are not the ones that the language model creates; instead, they are mapped to the corresponding vectors in the Word2Vec space [11]. Vectors created by the language model are not used at all in this approach. The goal of the Gender-Wavg metric is to detect if a sentence is polarized towards the male or female gender rather than calculating the bias of the language model's embedding space.

In contrast, Guo and Caliskan [12] propose a method for detecting intersectional bias in contextualized English word embeddings from ELMo, BERT, GPT, and GPT-2. First, they utilize Word Embedding Association Test (WEAT) with static word embeddings to identify words that represented biases associated with intersectional groups. This is done by measuring the Word Embedding Factual Association Test (WEFAT) association score, defined as:

$$\text{s}(\vec{w}, \mathcal{A}, \mathcal{B}) = \frac{\hat{\mathbb{E}}_{\vec{a} \in \mathcal{A}}[\cos(\vec{w}, \vec{a})] - \hat{\mathbb{E}}_{\vec{b} \in \mathcal{B}}[\cos(\vec{w}, \vec{b})]}{\hat{\mathbb{V}}_{\vec{x} \in \mathcal{A} \cap \mathcal{B}}[\cos(\vec{w}, \vec{x})]^{1/2}}, \tag{2}$$

where $\hat{\mathbb{E}}_{\vec{a} \in \mathcal{A}}$ and $\hat{\mathbb{V}}_{\vec{a} \in \mathcal{A}}$ represent, respectively, the empirical mean and empirical variance operators; $\mathcal{A}$ and $\mathcal{B}$ are sets of vectors encompassing concepts, e.g., male and female; $\vec{w} \in \mathcal{W}$; and $\mathcal{W}$ is a set of target stimuli, e.g., occupations. Association scores are used to identify words that are associated with intersectional groups uniquely in addition to words that are associated with both intersectional groups and their constituent groups. Once these words have been identified, the authors then extend WEAT to contextualized embeddings by calculating a distribution of effect sizes $\text{ES}(\mathcal{X}, \mathcal{Y}, \mathcal{A}, \mathcal{Y})$ among the sets of target words $\mathcal{X}$ and $\mathcal{Y}$, and the sets of concepts or attributes $\mathcal{A}$ and $\mathcal{B}$. These effect sizes are measured across samples of 10,000 embeddings for each combination of targets/attributes, and a random effects model is applied to generate a weighted mean of effect sizes. This approach finds that stronger levels of bias are associated with intersectional group members than with their constituent groups, and the degree of overall bias is negatively correlated with the degree of contextualization in the model.

2.3. *Bias Detection in Swedish Language Models*

Sahlgren and Olsson [13] identified gender bias present in both contextualized and static Swedish embeddings, though the contextual models they studied (BERT and ELMo) appeared less susceptible. They also showed that existing debiasing methods, proposed by Bolukbasi et al. [3], not only failed to mitigate bias in Swedish language models but possibly worsened existing stereotypes present in static embeddings. Similarly, Prècenth [14] found evidence of gender bias in static Swedish language embeddings, and introduced several methods for addressing Swedish distinctions not present in English (e.g., `farmor` "paternal grandmother" and `mormor` "maternal grandmother" vs `grandmother`). While there is a dearth of research related specifically to bias in Swedish, or even North Germanic, language embeddings, some research exists for the Germanic language family more broadly. Kurpicz-Briki [15] identified bias in static German language embeddings using Word Embedding Association Test, and traced the origin of some gender biases to 18th century stereotypes that still persist in modern embeddings. Matthews et al. [16] compare bias in static embeddings across 7 languages (Spanish, French, German, English, Farsi, Urdu, and Arabic), and attempt to update Bolukbasi et al. [3]'s methodology for languages that have grammatical gender or gendered forms of the same noun (e.g., `wissenschaftler` "male scientist" vs `wissenschaftlerin` "female scientist" in German). Additionally, Bartl et al. [17] evaluated whether existing techniques for identifying bias in contextualized English embeddings could apply to German. While they confirmed Kurita et al. [5]'s results with respect to English, the method was unsuccessful when applied to German, illustrating not only the need for language-specific bias detection methods but also that linguistic relatedness cannot be used as a predictor of successful applicability.

Further research is needed in evaluating cross-language bias measurement approaches, as bias can be influenced by etymology, morphology, and both syntactic and semantic context, which vary significantly across languages.

3. Methods

Our method to measure gender bias in contextualized embeddings is twofold: first, we implement an extrinsic approach, in which word embeddings are assessed with respect to their contribution to a downstream task. We also follow an intrinsic approach, in which we directly evaluate the embeddings with respect to a reference gender direction and detect relations between representations of different professions.

Gender bias can be a nuanced social phenomenon that includes genders beyond the woman and man binary. Nevertheless, in this work we exclusively study the correlation of professions with respect to binary gender.

3.1. *Extrinsic Evaluation of Gender Bias in T5 and mT5*

The downstream task used in this work is semantic text similarity. We use Text-to-Text Transfer Transformer (T5) and multilingual T5 (mT5), and we fine-tune two mT5 models: one on the English STS-B dataset (http://ixa2.si.ehu.es/stswiki/index.php/STSbenchmark, accessed on 12 December 2021) and one on the machine translated variant of it for Swedish [18]. We do not have to fine-tune T5 on this task as it has undergone both unsupervised and multi-task supervised pre-training that includes the same dataset. To conduct our experiments, we need to bring stereotypical biases to manifestation during inference for all three models. To this end, we create a new dataset by adapting the STS-B dataset.

3.1.1. Dataset Creation

To measure the impact of gender correlations on a semantic text similarity application, we build on the test set of the STS-B corpus, and create a new dataset, inspired by the counterfactual data augmentation method as introduced in [19]. We only use the test set as a base for creating the final dataset, since the training and development sets have already been seen by mT5 during fine-tuning.

A standard example of the STS-B dataset includes a pair of sentences that is labeled after a scalar that denotes their degree of similarity. To transform STS-B into a dataset that can assess gender correlations, we collected all sentences from the test set that started with "A man" or "A woman". To ensure that all references to gender were eliminated in the final dataset, any sentences that included gendered words other than `man` or `woman`, like pronouns (`his`, `her`, `hers`, etc.), were discarded; as a result `man` or `woman` were the only words present in each sample that could disclose gender information. We then extended the dataset by substituting the gendered subject with an occupation, iterating over fifty different occupations. The final dataset consists of 149 rows and 52 columns. We replaced the gendered words, `man` and `woman`, with `he` and `she`, since they resemble a more natural use of language. The same process is applied for the Swedish variant of the STS-B dataset.

3.1.2. Experimental Design

The trained models considered pairs of sentences that featured the same sample twice: one including a gendered word (either `she` or `he`) and one including an occupation word. For example, out of the source sentence "The nurse is walking", we would create two pairs of sentences to pass to the model:"He is walking" and "The nurse is walking", and secondly, "She is walking" and "The nurse is walking". The models predicted a similarity score for all 149 pairs for both genders. Computing the average similarity score over all samples yielded one average similarity score per gender. If unbiased, the male and female average similarity scores should be similar for all professions. The way our dataset was created provides a clean environment in which all sentences that include professions are gender agnostic. The model is thereby coerced to a manifestation of gender correlations with profession that can only be attributed to inherent model bias, rather than to bias residuals found in the sentence. This ensures the validity and reliability of this method. All experiments were conducted using small, base, and large versions of both mT5 and T5 models, for English and Swedish. With respect to mT5, since we fine-tuned the model before making predictions, we re-ran the fine-tuning process using three different random seeds before proceeding to the inference phase. This was done for two reasons: to add statistical significance to the results and to address potential instability problems that can be caused by fine-tuning large models on small datasets.

3.2. *Intrinsic Evaluation of Gender Bias in T5*

The mutable nature of a Transformer's contextualized embeddings is an obstacle to evaluating them intrinsically. Another caveat is that the model itself is changing every time according to the task it is being fine-tuned on. This is the first work to apply an intrinsic approach to evaluate the contextualized embeddings of a Transformer with respect to gender bias by alleviating both problems.

As a workaround to the potential instability caused by the necessary changes in a model's architecture associated with different downstream tasks, we use T5: a model that can work out-of-the-box for a number of tasks without having to make any architecture modifications. Nevertheless, mT5 was not pre-trained on many tasks the same way as T5. Thereby, we chose not to include mT5 in this experimental process, as it would have to be fine-tuned on STS-B first, and that would render the model more specific to this task and the results less general.

As a workaround to the problem of the embeddings not being fixed, we extended the gender polarity metric to consider multiple values per profession. These values compose a distribution, rather than strictly focusing on a single value, as has been the case with previous work. Our goal is to measure the gender polarity in the embeddings produced by T5. To this end, we were inspired by Dhamala et al. [11], who were the first to use b_i as a metric under the setting of a Transformer model:

$$b_i = \frac{\vec{w}_i \cdot \vec{g}}{\|\vec{w}_i\| \|\vec{g}\|}, \tag{3}$$

where $\vec{g} = \vec{she} - \vec{he}$. Nevertheless, the authors avoided the direct use of the contextual embeddings $\vec{w}_i$ when computing the bias b_i and chose to map them to the Word2Vec space first. The motivation behind their choice is that there was no theoretical foundation in literature to suggest that a constant gender direction, $\vec{g} = \vec{she} - \vec{he}$, exists in the embedding space of a Transformer model. Thus, they settled for the fixed embedding space of Word2Vec which can safely establish a well defined $\vec{g}$.

In this work, we make the hypothesis that a versatile Transformer model like T5, which already holds the knowledge of various downstream tasks due to the multi-task pre-training procedure it has undergone, can still establish a gender direction, $\vec{g} = \vec{she} - \vec{he}$. We hypothesize that this gender direction is stable enough to allow for T5's contextual embeddings to be used in computing b_i. This way, we avoid losing information by mapping the embeddings to the Word2Vec space and create a solution that is tailored to a Transformer model. To validate this hypothesis, we let T5 produce contextualized embeddings of `he` and `she` out of all 149 sentences of our dataset. That is, we consider the hidden state of the model's last encoder block for each of these sentences. We used the small, base, and large version of T5. Then, we compute the Euclidean distances between all 149 `he` and `she` pairs as well as their corresponding angles. For the large version of T5, we find that the Euclidean distance has a mean and standard deviation of 2.79 ± 0.22 and the angle has a mean and standard deviation of 0.68 ± 0.04 radians. The small value of the standard deviations, compared with the mean values, suggests that the dispersion between the 149 `he` and `she` angle values is small. This indicates that there might exist a well defined, and perhaps constant, gender direction $\vec{g}$ between `he` and `she` in the T5 embedding space. We use the average vector $\vec{g} = \frac{1}{149}\sum_{i=1}^{149}(\vec{she}_i - \vec{he}_i)$ as the gender direction and compute gender polarity b_i for '`he` and `she`, and nine selected occupations: `nurse`, `engineer`, `surgeon`, `scientist`, `receptionist`, `programmer`, `teacher`, `officer`, and `homemaker`. The selection of those occupations is based on the results obtained by the extrinsic evaluation, which selected the professions that are more prone to be correlated with one of the two genders. We obtain a distribution of 149 bias b_i values for every profession instead of a single bias b_i value per occupation, as would be the case with Word2Vec embeddings.

4. Results

4.1. Extrinsic Evaluation of Gender Bias in T5 and mT5

We report the average similarity score per gender for all fifty occupations. Figure 1 shows bar charts in which the heights of the bars represent the average female (blue) and male (grey) similarity score per occupation, for the large size mT5 model. Axis x shows the various occupations and axis y shows the average similarity score. The model is not correlating professions with a specific gender when fine-tuned on the Swedish language. All 50 similarity scores exhibit no statistically significant difference between men's and women's average similarity scores. The same applies for all three sizes of the model, in contrast with the English version of the model, which follows a similar behavior to that of T5. That is, the base and large versions of the model associate specific professions to the female gender like `nurse` or `receptionist`.

Figure 2 presents the average difference between mean similarity scores for men and women over the 50 occupations. Mean differences tend to grow larger for larger sizes of the model. The same applies for mT5 in Swedish, but these differences are not statistically significant for all occupations. Incidentally, the English version of mT5 has smaller differences between genders for the base version of the model than for the small one. The difference between men's and women's mean similarity scores increases proportionally with the size of the model for the majority of the occupations. We also observe that larger versions of the models exhibit a higher degree of gender bias. Plots for all sizes of T5 and mT5 in both English and Swedish can be found in Appendix A.

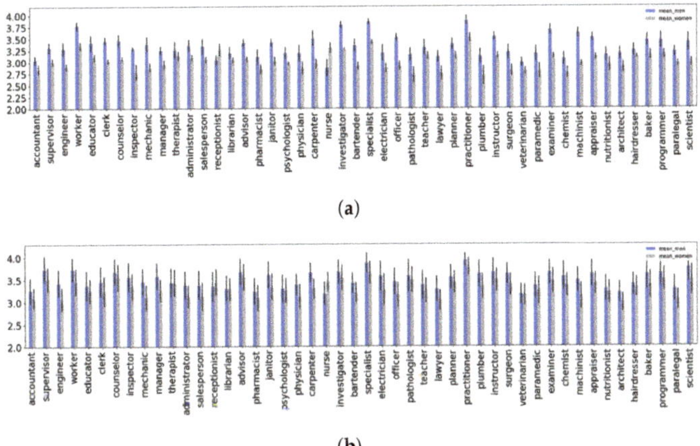

Figure 1. (**a**) Average similarity scores per occupation. Language: English, (**b**) Average similarity scores per occupation. Language: Swedish. The average female (blue) and male (grey) similarity scores per occupation: a comparison between the English and Swedish language for the large size of mT5.

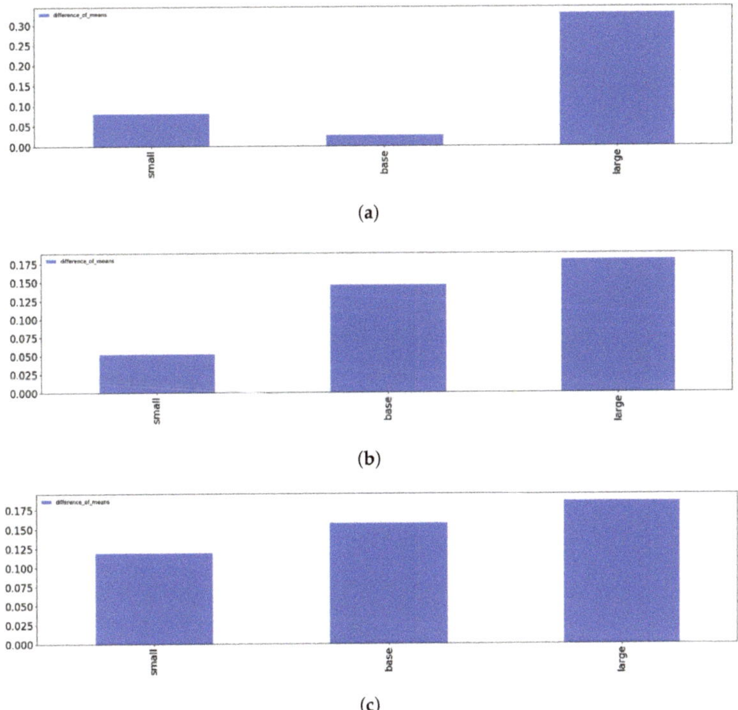

Figure 2. (**a**) Mean difference between gender similarity scores per model size. Model: mT5. Language: English. (**b**) Mean difference between gender similarity scores per model size. Model: mT5. Language: Swedish. (**c**) Mean difference between gender similarity scores per model size. Model: T5. Language: English. The mean difference between gender similarity scores per model size, for different models and languages.

4.2. Intrinsic Evaluation of Gender Bias in T5

Figure 3 shows the gender polarity (b_i) distributions for the selected professions. Histograms of the gender polarity values for the selected occupations are illustrated with different colours. The graph compares the three different sizes of T5. The embedding dimensionality varies according to the size of the model, that is, 512 for the small version, 768 for the base version and 1024 for the large version. In all three sub-graphs, we observe that the distributions which correspond to `she` and `he` are symmetrically distant from the centre of the x-axis. Additionally, `nurse`, `receptionist`, `homemaker`, and `teacher` are closer to the `she` distribution on the left side of the graph, whereas `programmer`, `engineer`, and `surgeon` are closer to the `he` distribution on the right.

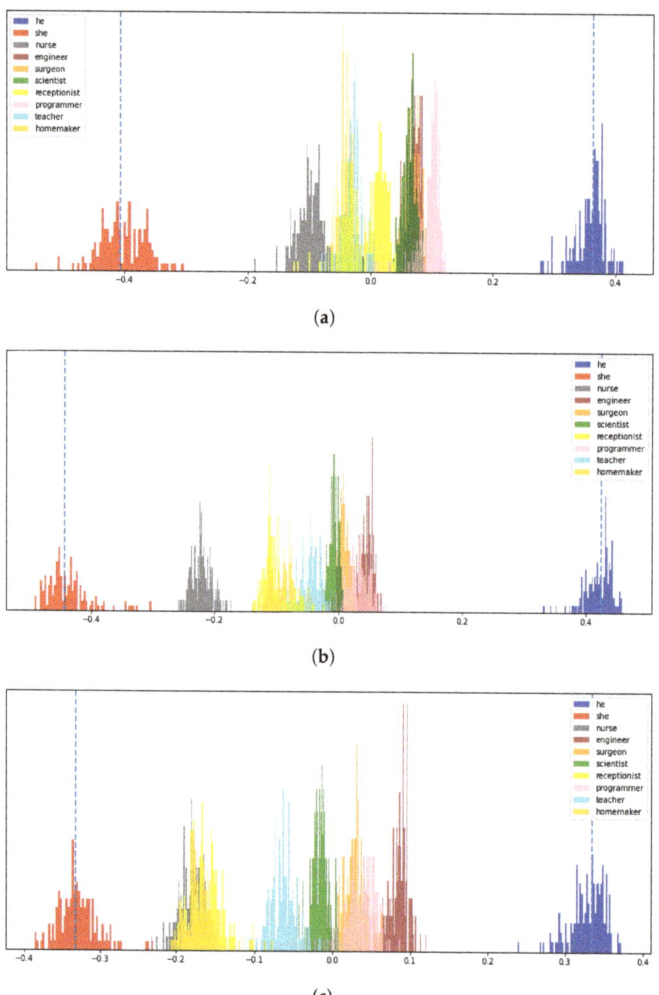

Figure 3. (**a**) The 149 b_i values per occupation for the small size of T5. Embedding dimensionality: 512. (**b**) The 149 b_i values per occupation for the base size of T5. Embedding dimensionality: 768. (**c**) The 149 b_i values per occupation for the large size of T5. Embedding dimensionality: 1024. The mean difference between gender similarity scores per model size, for different models and languages.

By comparing all three sub-graphs in Figure 3, we notice that the gulf between the various occupation distributions grows larger as the model's size increases; there is a high overlap of the distributions for the small size of T5, which indicates that the occupations are less gender polarized. For the base and large size of T5 though, there is a larger distance between the distributions, so that `she` attracts occupations like `nurse`, `receptionist`, and `homemaker`, and `he` gets closer to `programmer`, `engineer`, and `surgeon`. Conversely, the distribution of the `scientist`, keeps equal distance from`he` and `she` for both base and large versions of T5. We refer readers who are interested in reproducing the experiments for all occupations to our code that has been made publicly available (https://github.com/Stellakats/Master-thesis-gender-bias, accessed on 12 December 2021).

5. Discussion

In this paper, we introduced an intrinsic approach to measuring gender bias on contextualized embeddings by using gender polarity: an existing bias metric that measures how related an embedding of a word is to a specific gender. This metric has previously been applied on contextualized embeddings by first mapping them to the Word2Vec embedding space. We contribute by first detecting a stable gender direction in T5's embedding space and then computing gender polarity distributions for the various embeddings, instead of single values, for each word. The results of this approach are consistent with those of an extrinsic approach that we also followed; we evaluated T5's and mT5's outputs in terms of how bias can be propagated to the downstream task of semantic text similarity.

Our results indicate that higher status professions tend to be more associated to the male gender than the female gender. We also compared Swedish with English as well as various model sizes and found that our methods find less bias associated with gender in the Swedish language, though we note that the detection method itself may be more sensitive to bias in English. Additionally, we find that larger sizes of the models can lead to an increased manifestation of gender bias. This finding suggests that the embedding dimensionality might be proportional to the extend to which biases will be successfully encoded in the embedding vectors.

The consistency of the results between the intrinsic and extrinsic approach might be a positive indicator that deriving a stable gender direction in a Transformer model's embedding space is feasible and can lead to valid results. This is a simple, yet powerful idea, which if supported by further research, can offer a solid basis for effective debiasing in Transformer models.

6. Ethics Statement

It has been shown that changes in stereotypes and attitudes towards women and their participation in the workforce can be quantified by tracking the temporal dynamics of bias in word embeddings [20]. Furthermore, it has been observed in various use cases that models might marginalize specific groups in the way they handle downstream tasks, establishing a behavior similar to that of a stereotypically biased conduct [21–26]. To responsibly direct actions that will combat this problem, it is of crucial importance that we find reliable ways of detecting and quantifying it, which is what we aim for in this work. A reliable way of bias detection could be the touchstone of developing effective bias mitigation techniques, which could practically contribute to the pursuit of a more fair representation of different races and genders by the models. Such a course of action complies with the fifth and tenth goal regarding "gender equality" and "reduced inequalities" respectively, as defined in the 17 Sustainable Development Goals (https://sdgs.un.org/goals, accessed on 12 December 2021) set by the United Nations General Assembly and intended to be achieved by the year 2030. More specifically, this work is aligned with sub-goal 10.2 that is about empowering and promoting "the social, economic and political inclusion of all, irrespective of age, sex, disability, race, ethnicity, origin, religion or economic or other status" (https://sdgs.un.org/goals/goal10, accessed on 12 December 2021). This work is also aligned with sub-goal 5.1 that is about ending "all forms of discrimination against women and girls everywhere", and sub-goal 5.5, which ensures "women's full and effective participation and equal opportunities for leadership at all levels of decision-making in political, economic and public life" (https://sdgs.un.org/goals/goal5, accessed on 12 December 2021).

Author Contributions: Conceptualization, S.K. and B.R.-G.; methodology, S.K. and B.R.-G.; software, S.K.; validation, S.K., B.R.-G. and J.S.; formal analysis, S.K.; investigation, S.K.; data curation, S.K.; writing—original draft preparation, S.K.; writing—review and editing, S.K., B.R.-G. and J.S.; visualization, S.K., B.R.-G. and J.S.; supervision, B.R.-G.; project administration, S.K. and B.R.-G.; All authors have read and agreed to the published version of the manuscript.

Funding: This research was funded in part by by VINNOVA, Sweden (project title: Språkmodeller för svenska myndigheter, grant number 2019-02996) and by the Swedish research council under contract 2019-03606.

Institutional Review Board Statement: Not applicable.

Informed Consent Statement: Not applicable.

Data Availability Statement: The data that support the findings of this study are openly available at http://ixa2.si.ehu.es/stswiki/index.php/STSbenchmark accessed on 12 December 2021, at https://github.com/timpal0l/sts-benchmark-swedish accessed on 12 December 2021 and at https://github.com/Stellakats/Master-thesis-gender-bias accessed on 12 December 2021.

Conflicts of Interest: The authors declare no conflict of interest.

Appendix A

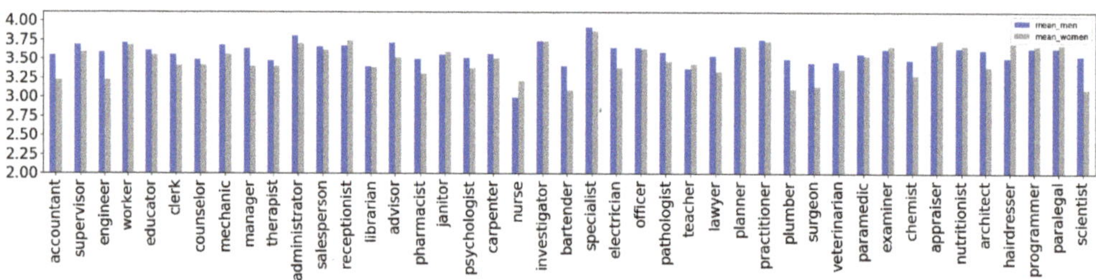

Figure A1. Average similarity scores per occupation. Model: T5 small.

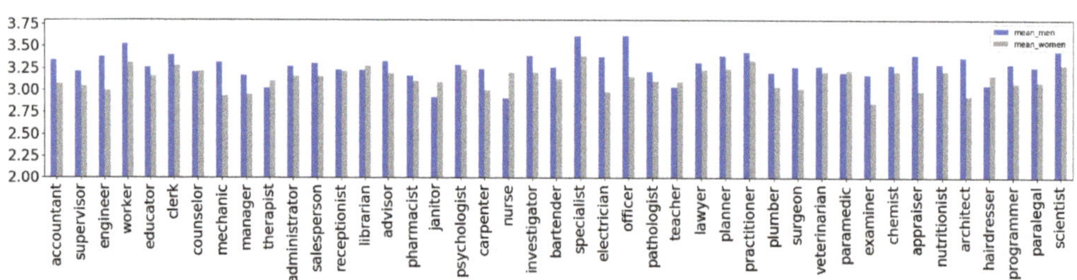

Figure A2. Average similarity scores per occupation. Model: T5 base.

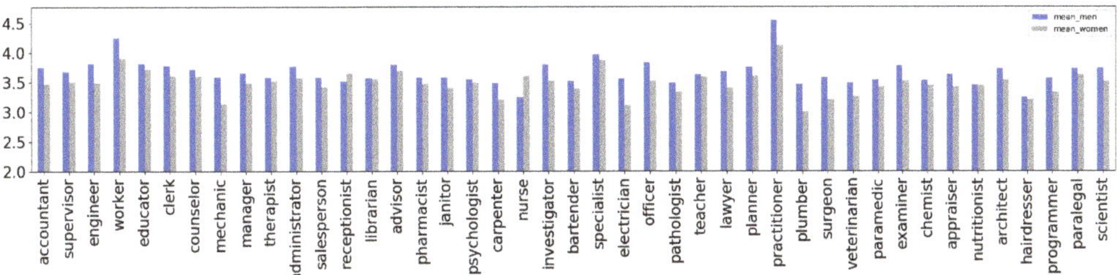

Figure A3. Average similarity scores per occupation. Model: T5 large.

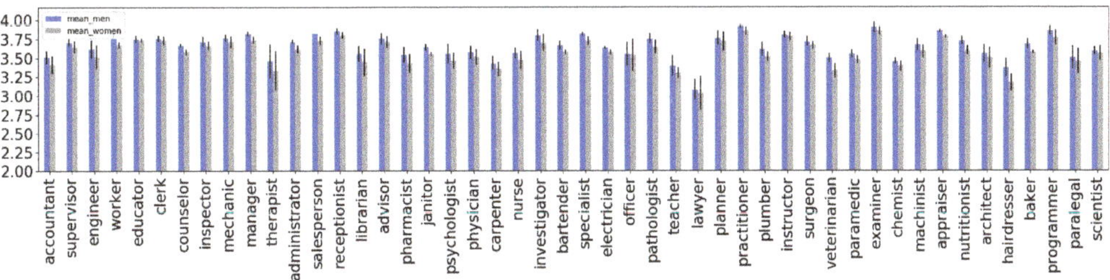

Figure A4. Average similarity scores per occupation. Language: English. Model: mT5 small.

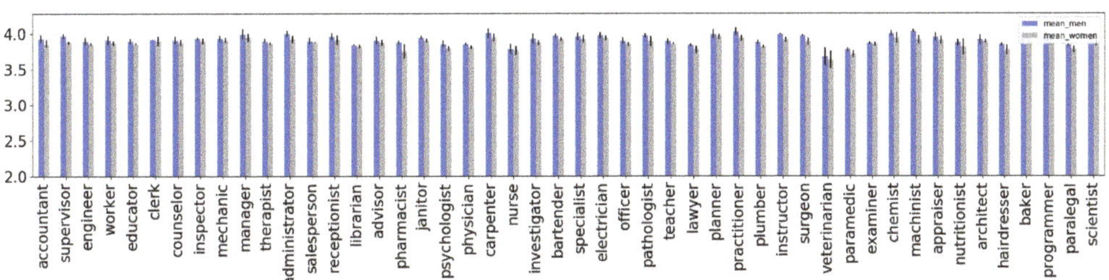

Figure A5. Average similarity scores per occupation. Language: Swedish. Model: mT5 small.

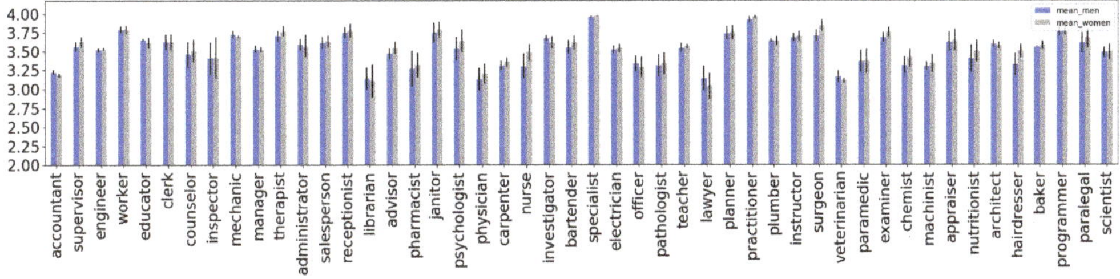

Figure A6. Average similarity scores per occupation. Language: English. Model: mT5 base.

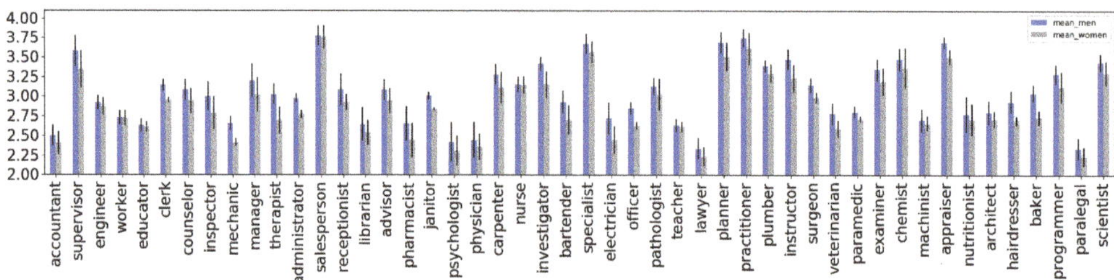

Figure A7. Average similarity scores per occupation. Language: Swedish. Model size: mT5 base.

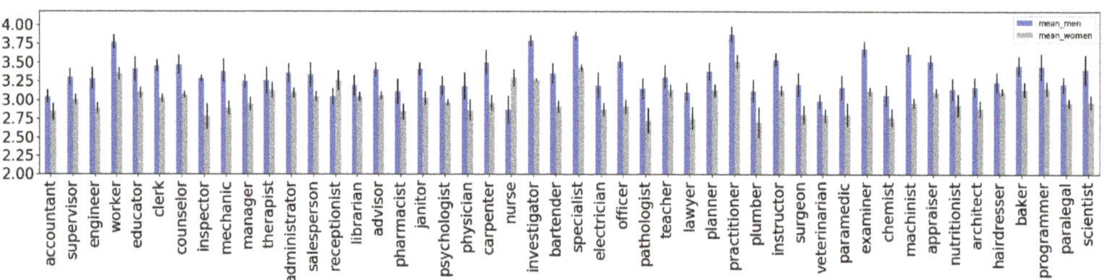

Figure A8. Average similarity scores per occupation. Language: English. Model: mT5 large.

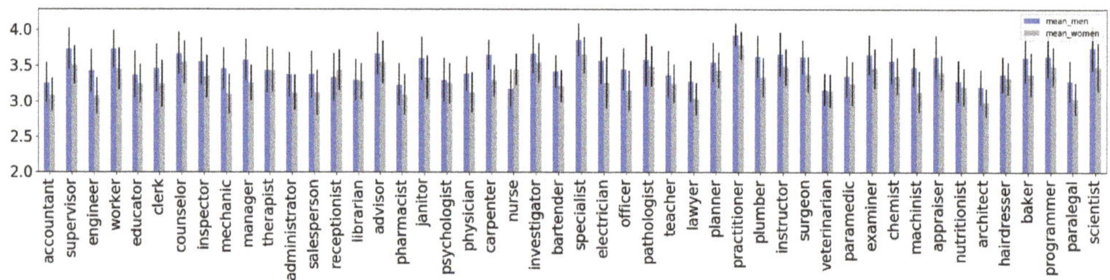

Figure A9. Average similarity scores per occupation. Language: Swedish. Model: mT5 large.

References

1. Mikolov, T.; Chen, K.; Corrado, G.; Dean, J. Efficient Estimation of Word Representations in Vector Space. In Proceedings of the 1st International Conference on Learning Representations, ICLR 2013, Scottsdale, AZ, USA, 2–4 May 2013.
2. Zhao, J.; Zhou, Y.; Li, Z.; Wang, W.; Chang, K.W. Learning Gender-Neutral Word Embeddings. In Proceedings of the 2018 Conference on Empirical Methods in Natural Language Processing, Brussels, Belgium, 31 October–4 November 2018; pp. 4847–4853. [CrossRef]
3. Bolukbasi, T.; Chang, K.W.; Zou, J.Y.; Saligrama, V.; Kalai, A.T. Man is to computer programmer as woman is to homemaker? debiasing word embeddings. *Adv. Neural Inf. Process. Syst.* **2016**, *26*, 4349–4357.
4. Burstein, J.; Doran, C.; Solorio, T. *Proceedings of the 2019 Conference of the North American Chapter of the Association for Computational Linguistics: Human Language Technologies, NAACL-HLT 2019, Minneapolis, MN, USA, 2–7 June 2019, Volume 1 (Long and Short Papers)*; Association for Computational Linguistic: Minneapolis, MN, USA, 2019.
5. Kurita, K.; Vyas, N.; Pareek, A.; Black, A.; Tsvetkov, Y. Measuring Bias in Contextualized Word Representations. In Proceedings of the First Workshop on Gender Bias in Natural Language Processing, Florence, Italy, 2 August 2019; pp. 166–172. [CrossRef]
6. Caliskan, A.; Bryson, J.J.; Narayanan, A. Semantics derived automatically from language corpora contain human-like biases. *Science* **2017**, *356*, 183–186. [CrossRef] [PubMed]
7. Devlin, J.; Chang, M.W.; Lee, K.; Toutanova, K. BERT: Pre-training of Deep Bidirectional Transformers for Language Understanding. *arXiv* **2019**, arXiv:1810.04805.
8. Nangia, N.; Vania, C.; Bhalerao, R.; Bowman, S.R. CrowS-Pairs: A Challenge Dataset for Measuring Social Biases in Masked Language Models. *arXiv* **2020**, arXiv:2010.00133.
9. Nadeem, M.; Bethke, A.; Reddy, S. StereoSet: Measuring stereotypical bias in pretrained language models. *arXiv* **2020**, arXiv:2004.09456.

10. Webster, K.; Wang, X.; Tenney, I.; Beutel, A.; Pitler, E.; Pavlick, E.; Chen, J.; Petrov, S. Measuring and Reducing Gendered Correlations in Pre-trained Models. *arXiv* **2020**, arXiv:2010.06032.
11. Dhamala, J.; Sun, T.; Kumar, V.; Krishna, S.; Pruksachatkun, Y.; Chang, K.W.; Gupta, R. Bold: Dataset and metrics for measuring biases in open-ended language generation. In Proceedings of the 2021 ACM Conference on Fairness, Accountability, and Transparency, Virtual Event, 3–10 March 2021; pp. 862–872.
12. Guo, W.; Caliskan, A. Detecting Emergent Intersectional Biases: Contextualized Word Embeddings Contain a Distribution of Human-like Biases. In Proceedings of the 2021 AAAI/ACM Conference on AI, Ethics, and Society, Virtual Event, 19–21 May 2021; Association for Computing Machinery: New York, NY, USA, 2021; pp. 122–133. [CrossRef]
13. Sahlgren, M.; Olsson, F. Gender Bias in Pretrained Swedish Embeddings. In Proceedings of the 22nd Nordic Conference on Computational Linguistics, Turku, Finland, 30 September–2 October 2019; Linköping University Electronic Press: Linköping, Sweden; pp. 35–43.
14. Prècenth, R. Word Embeddings and Gender Stereotypes in Swedish and English. Ph.D. Thesis, Uppsala University, Uppsala, Sweden, 2019.
15. Kurpicz-Briki, M. Cultural Differences in Bias? Origin and Gender Bias in Pre-Trained German and French Word Embeddings. In Proceedings of the 5th Swiss Text Analytics Conference and the 16th Conference on Natural Language Processing, SwissText/KONVENS 2020, Online, 23–25 June 2020; Ebling, S., Tuggener, D., Hürlimann, M., Cieliebak, M., Volk, M., Eds.; CEUR Workshop Proceedings: Zurich, Switzerland, 2020; Volume 2624.
16. Matthews, A.; Grasso, I.; Mahoney, C.; Chen, Y.; Wali, E.; Middleton, T.; Njie, M.; Matthews, J. Gender Bias in Natural Language Processing Across Human Languages. In Proceedings of the First Workshop on Trustworthy Natural Language Processing, Online, 10 June 2021; Association for Computational Linguistics: Barcelona, Spain, 2021; pp. 45–54. [CrossRef]
17. Bartl, M.; Nissim, M.; Gatt, A. Unmasking Contextual Stereotypes: Measuring and Mitigating BERT's Gender Bias. In Proceedings of the Second Workshop on Gender Bias in Natural Language Processing, Barcelona, Spain, 13 December 2020; Association for Computational Linguistics: Stroudsburg, PA, USA, 2020; pp. 1–16.
18. Isbister, T.; Sahlgren, M. Why Not Simply Translate? A First Swedish Evaluation Benchmark for Semantic Similarity. *arXiv* **2020**, arXiv:2009.03116.
19. Lu, K.; Mardziel, P.; Wu, F.; Amancharla, P.; Datta, A. Gender Bias in Neural Natural Language Processing. *arXiv* **2019**, arXiv:1807.11714.
20. Garg, N.; Schiebinger, L.; Jurafsky, D.; Zou, J. Word embeddings quantify 100 years of gender and ethnic stereotypes. *Proc. Natl. Acad. Sci. USA* **2018**, *115*, E3635–E3644. [CrossRef] [PubMed]
21. Basta, C.; Costa-jussà, M.R.; Casas, N. Evaluating the Underlying Gender Bias in Contextualized Word Embeddings. In Proceedings of the First Workshop on Gender Bias in Natural Language Processing, Florence, Italy, 2 August 2019; Association for Computational Linguistics: Stroudsburg, PA, USA, 2019; pp. 33–39. [CrossRef]
22. Beltagy, I.; Lo, K.; Cohan, A. SciBERT: A Pretrained Language Model for Scientific Text. In Proceedings of the 2019 Conference on Empirical Methods in Natural Language Processing and the 9th International Joint Conference on Natural Language Processing (EMNLP-IJCNLP), Hong Kong, China, 3–7 November 2019; Association for Computational Linguistics: Stroudsburg, PA, USA; pp. 3615–3620. [CrossRef]
23. Hutchinson, B.; Prabhakaran, V.; Denton, E.; Webster, K.; Zhong, Y.; Denuyl, S.C. Social Biases in NLP Models as Barriers for Persons with Disabilities. In Proceedings of the 58th Annual Meeting of the Association for Computational Linguistics, Online, 5–10 July 2020; Association for Computational Linguistics: Stroudsburg, PA, USA, 2020; pp. 5491–5501.
24. Sheng, E.; Chang, K.W.; Natarajan, P.; Peng, N. The woman worked as a babysitter: On biases in language generation. *arXiv* **2019**, arXiv:1909.01326.
25. Zhang, H.; Lu, A.X.; Abdalla, M.; McDermott, M.; Ghassemi, M. Hurtful words: Quantifying biases in clinical contextual word embeddings. In Proceedings of the ACM Conference on Health, Inference, and Learning, Toronto, ON, Canada, 2–4 April 2020; Association for Computing Machinery: New York, NY, USA, 2020; pp. 110–120.
26. Zhou, P.; Shi, W.; Zhao, J.; Huang, K.H.; Chen, M.; Cotterell, R.; Chang, K.W. Examining Gender Bias in Languages with Grammatical Gender. In Proceedings of the 2019 Conference on Empirical Methods in Natural Language Processing and the 9th International Joint Conference on Natural Language Processing (EMNLP-IJCNLP), Hong Kong, China, 3–7 November 2019; Association for Computational Linguistics: Stroudsburg, PA, USA, 2019; pp. 5276–5284.

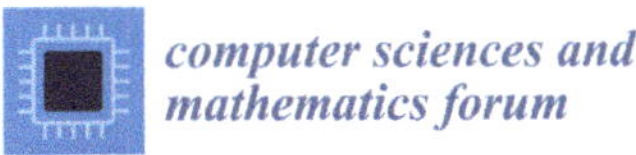
computer sciences and mathematics forum

Proceeding Paper

The Details Matter: Preventing Class Collapse in Supervised Contrastive Learning †

Daniel Y. Fu *,‡, Mayee F. Chen ‡, Michael Zhang, Kayvon Fatahalian and Christopher Ré

Department of Computer Science, Stanford University, Stanford, CA 94035, USA;
mfchen@cs.stanford.edu (M.F.C.); mzhang@cs.stanford.edu (M.Z.); kayvonf@cs.stanford.edu (K.F.);
chrismre@cs.stanford.edu (C.R.)
* Correspondence: danfu@cs.stanford.edu
† Presented at the AAAI Workshop on Artificial Intelligence with Biased or Scarce Data (AIBSD), Online, 28 February 2022.
‡ These authors contributed equally to this work.

Abstract: Supervised contrastive learning optimizes a loss that pushes together embeddings of points from the same class while pulling apart embeddings of points from different classes. Class collapse—when every point from the same class has the same embedding—minimizes this loss but loses critical information that is not encoded in the class labels. For instance, the "cat" label does not capture unlabeled categories such as breeds, poses, or backgrounds (which we call "strata"). As a result, class collapse produces embeddings that are less useful for downstream applications such as transfer learning and achieves suboptimal generalization error when there are strata. We explore a simple modification to supervised contrastive loss that aims to prevent class collapse by uniformly pulling apart individual points from the same class. We seek to understand the effects of this loss by examining how it embeds strata of different sizes, finding that it clusters larger strata more tightly than smaller strata. As a result, our loss function produces embeddings that better distinguish strata in embedding space, which produces lift on three downstream applications: 4.4 points on coarse-to-fine transfer learning, 2.5 points on worst-group robustness, and 1.0 points on minimal coreset construction. Our loss also produces more accurate models, with up to 4.0 points of lift across 9 tasks.

Keywords: contrastive learning; supervised contrastive learning; transfer learning; robustness; noisy labels; coresets

Citation: Fu, D.Y.; Chen, M.F.; Zhang, M.; Fatahalian, K.; Ré, C. The Details Matter: Preventing Class Collapse in Supervised Contrastive Learning. *CSFM* **2022**, 3, 4. https://doi.org/10.3390/cmsf2022003004

Academic Editors: Kuan-Chuan Peng and Ziyan Wu

Published: 15 April 2022

Publisher's Note: MDPI stays neutral with regard to jurisdictional claims in published maps and institutional affiliations.

1. Introduction

Supervised contrastive learning has emerged as a promising method for training deep models, with strong empirical results over traditional supervised learning [1]. Recent theoretical work has shown that under certain assumptions, *class collapse*—when the representation of every point from a class collapses to the same embedding on the hypersphere, as in Figure 1—minimizes the supervised contrastive loss L_{SC} [2]. Furthermore, modern deep networks, which can memorize arbitrary labels [3], are powerful enough to produce class collapse.

Although class collapse minimizes L_{SC} and produces accurate models, it loses information that is not explicitly encoded in the class labels. For example, consider images with the label "cat." As shown in Figure 1, some cats may be sleeping, some may be jumping, and some may be swatting at a bug. We call each of these semantically-unique categories of data—some of which are rarer than others, and none of which are explicitly labeled—a *stratum*. Distinguishing strata is important; it empirically can improve model performance [4] and fine-grained robustness [5]. It is also critical in high-stakes applications such as medical imaging [6]. However, L_{SC} maps the sleeping, jumping, and swatting cats all to a single "cat" embedding, losing strata information. As a result, these embeddings are less useful for common downstream applications in the modern machine learning landscape, such as transfer learning.

In this paper, we explore a simple modification to L_{SC} that prevents class collapse. We study how this modification affects embedding quality by considering how strata are represented in embedding space. We evaluate our loss both in terms of embedding quality, which we evaluate through three downstream applications, and end model quality.

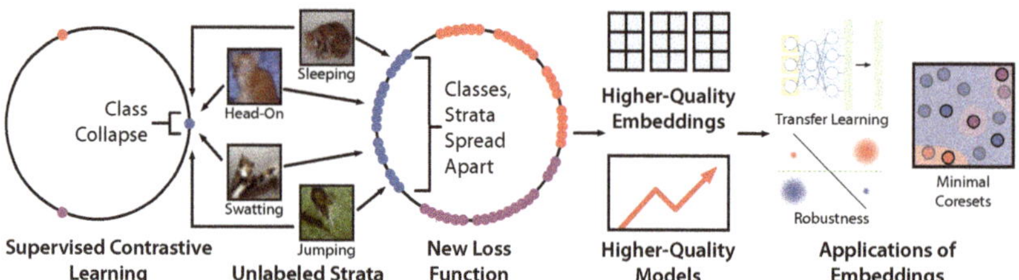

Figure 1. Classes contain critical information that is not explicitly encoded in the class labels. Supervised contrastive learning (left) loses this information, since it maps unlabeled strata such as sleeping cats, jumping cats, and swatting cat to a single embedding. We introduce a new loss function L_{spread} that prevents class collapse and maintains strata distinctions. L_{spread} produces higher-quality embeddings, which we evaluate with three downstream applications.

In Section 3, we present our modification to L_{SC}, which prevents class collapse by changing how embeddings are pushed and pulled apart. L_{SC} pushes together embeddings of points from the same class and pulls apart embeddings of points from different classes. In contrast, our modified loss L_{spread} includes an additional class-conditional InfoNCE loss term that uniformly pulls apart individual points from within the same class. This term on its own encourages points from the same class to be maximally spread apart in embedding space, which discourages class collapse (see Figure 1 middle). Even though L_{spread} does not use strata labels, we observe that it still produces embeddings that qualitatively appear to retain more strata information than those produced by L_{SC} (see Figure 2).

In Section 4, motivated by these empirical observations, we study how well L_{spread} preserves distinctions between strata in the representation space. Previous theoretical tools that study the optimal embedding distribution fail to characterize the geometry of strata. Instead, we propose a simple thought experiment considering the embeddings that the supervised contrastive loss generates when it is trained on a partial sample of the dataset. This setup enables us to distinguish strata based on their sizes by considering how likely it is for them to be represented in the sample (larger strata are more likely to appear in a small sample). In particular, we find that points from rarer and more distinct strata are clustered less tightly than points from common strata, and we show that this clustering property can improve embedding quality and generalization error.

In Section 5, we empirically validate several downstream implications of these insights. First, we demonstrate that L_{spread} produces embeddings that retain more information about strata, resulting in lift on three downstream applications that require strata recovery:

- We evaluate how well L_{spread}'s embeddings encode fine-grained subclasses with coarse-to-fine transfer learning. L_{spread} achieves up to 4.4 points of lift across four datasets.
- We evaluate how well embeddings produced by L_{spread} can recover strata in an unsupervised setting by evaluating robustness against worst-group accuracy and noisy labels. We use our insights about how L_{spread} embeds strata of different sizes to improve worst-group robustness by up to 2.5 points and to recover 75% performance when 20% of the labels are noisy.
- We evaluate how well we can differentiate rare strata from common strata by constructing limited subsets of the training data that can achieve the highest performance under a fixed training strategy (the coreset problem). We construct coresets by subsampling

points from common strata. Our coresets outperform prior work by 1.0 points when coreset size is 30% of the training set.

Next, we find that L_{spread} produces higher-quality models, outperforming L_{SC} by up to 4.0 points across 9 tasks. Finally, we discuss related work in Section 6 and conclude in Section 7.

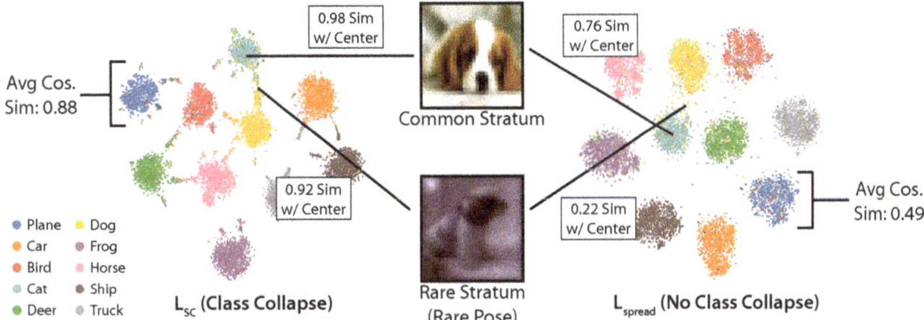

Figure 2. L_{spread} produces embeddings that are qualitatively better than those produced by L_{SC}. We show t-SNE visualizations of embeddings for the CIFAR10 test set and report cosine similarity metrics (average intracluster cosine similarities, and similarities between individual points and the class cluster). L_{spread} produces lower intraclass cosine similarity and embeds images from rare strata further out over the hypersphere than L_{SC}.

2. Background

We present our generative model for strata (Section 2.1). Then, we discuss supervised contrastive learning—in particular the SupCon loss L_{SC} from [1] and its optimal embedding distribution [2]—and the end model for classification (Section 2.2).

2.1. Data Setup

We have a labeled input dataset $\mathcal{D} = \{(x_i, y_i)\}_{i=1}^N$, where $(x, y) \sim \mathcal{P}$ for $x \in \mathcal{X}$ and $y \in \mathcal{Y} = \{1, \ldots, K\}$. For a particular data point x, we denote its label as $h(x) \in \mathcal{Y}$ with distribution $p(y|x)$. We assume that data is class-balanced such that $p(y = i) = \frac{1}{K}$ for all $i \in \mathcal{Y}$. The goal is to learn a model $\hat{p}(y|x)$ on $\mathcal{D}$ to classify points.

Data points also belong to categories beyond their labels, called *strata*. Following [5], we denote a stratum as a latent variable z, which can take on values in $\mathcal{Z} = \{1, \ldots, C\}$. $\mathcal{Z}$ can be partitioned into disjoint subsets $S_1, \ldots, S_K$ such that if $z \in S_k$, then its corresponding y label is equal to k. Let $S(c)$ denote the deterministic label corresponding to stratum c. We model the data generating process as follows. First, the latent stratum is sampled from distribution $p(z)$. Then, the data point x is sampled from the distribution $\mathcal{P}_z = p(\cdot|z)$, and its corresponding label is $y = S(z)$ (see Figure 2 of [5]). We assume that each class has m strata, and that there exist at least two strata, z_1, z_2, where $S(z_1) \neq S(z_2)$ and $\text{supp}(z_1) \cap \text{supp}(z_2) \neq \emptyset$.

2.2. Supervised Contrastive Loss

Supervised contrastive loss pushes together pairs of points from the same class (called positives) and pulls apart pairs of points from different classes (called negatives) to train an encoder $f : \mathcal{X} \to \mathbb{R}^d$. Following previous works, we make three assumptions on the encoder: (1) we restrict the encoder output space to be $\mathbb{S}^{d-1}$, the unit hypersphere; (2) we assume $K \leq d + 1$, which allows Graf et al. [2] to recover optimal embedding geometry; and (3) we assume the encoder f is "infinitely powerful", meaning that any distribution on $\mathbb{S}^{d-1}$ is realizable by $f(x)$.

2.2.1. SupCon and Collapsed Embeddings

We focus on the SupCon loss L_{SC} from [1]. Denote $\sigma(x, x') = f(x)^\top f(x')/\tau$, where τ is a temperature hyperparameter. Let $\mathcal{B}$ be the set of batches of labeled data on $\mathcal{D}$ and

$P(i,B) = \{p \in B\backslash i : h(p) = h(i)\}$ be the points in B with the same label as x_i. For an anchor x_i, the SupCon loss is $\hat{L}_{SC}(f, x_i, B) = \frac{-1}{|P(i,B)|} \sum_{p \in P(i,B)} \log \frac{\exp(\sigma(x_i, x_p))}{\sum_{a \in B\backslash i} \exp(\sigma(x_i, x_a))}$, where $P(i,B)$ forms positive pairs and $B\backslash i$ forms negative pairs.

The optimal embedding distribution that minimizes L_{SC} has one embedding per class, with the per-class embeddings collectively forming a regular simplex inscribed in the hypersphere Graf et al. [2]. Formally, if $h(x) = i$, then $f(x) = v_i$ for all $x \in \mathcal{B}$. $\{v_i\}_{i=1}^{K}$ makes up the regular simplex, defined by: a) $\sum_{i=1}^{K} v_i = 0$; b) $\|v_i\|_2 = 1$; and c) $\exists c_K \in \mathbb{R}$ s.t. $v_i^\top v_j = c_K$ for $i \neq j$. We describe this property as *class collapse* and define the distribution of $f(x)$ that satisfies these conditions as *collapsed embeddings*.

2.2.2. End Model

After the supervised contrastive loss is used to train an encoder, a linear classifier $W \in \mathbb{R}^{K \times d}$ is trained on top of the representations $f(x)$ by minimizing cross-entropy loss over softmax scores. We assume that $\|W_y\|_2 \leq 1$ for each $y \in \mathcal{Y}$. The end model's empirical loss can be defined as $\hat{\mathcal{L}}(W, \mathcal{D}) = \sum_{x_i \in \mathcal{D}} -\log \frac{\exp(f(x_i)^\top W_{h(x_i)})}{\sum_{j=1}^{K} \exp(f(x_i)^\top W_j)}$. The model uses softmax scores constructed with $f(x)$ and W to generate predictions $\hat{p}(y|x)$, which we also write as $\hat{p}(y|f(x))$. Finally, the generalization error of the model on $\mathcal{P}$ is the expected cross-entropy between $\hat{p}(y|x)$ and $p(y|x)$, namely $\mathcal{L}(x, y, f) = \mathbb{E}_{x,y \sim \mathcal{P}}[-\log \hat{p}(y|f(x))]$.

3. Method

We now highlight some theoretical problems with class collapse under our generative model of strata (Section 3.1). We then propose and qualitatively analyze the loss function L_{spread} (Section 3.2).

3.1. Theoretical Motivation

We show that the conditions under which collapsed embeddings minimize generalization error on coarse-to-fine transfer and the original task do *not* hold when distinct strata exist.

Consider the downstream *coarse-to-fine transfer* task (x, z) of using embeddings $f(x)$ learned on (x, y) to classify points by fine-grained strata. Formally, coarse-to-fine transfer involves learning an end model with weight matrix $W \in \mathbb{R}^{C \times d}$ and fixed $f(x)$ (as described in Section 2.2) on points (x, z), where we assume the data are class-balanced across z.

Observation 1. *Class collapse minimizes $\mathcal{L}(x, z, f)$ if for all x, (1) $p(y = h(x)|x) = 1$, meaning that each x is deterministically assigned to one class, and (2) $p(z|x) = \frac{1}{m}$ where $z \in S_{h(x)}$. The second condition implies that $p(x|z) = p(x|y)$ for all $z \in S_y$, meaning that there is no distinction among strata from the same class. This contradicts our data model described in Section 2.1.*

Similarly, we characterize when collapsed embeddings are optimal for the original task (x, y).

Observation 2. *Class collapse minimizes $\mathcal{L}(x, y, f)$ if, for all x, $p(y = h(x)|x) = 1$. This contradicts our data model.*

Proofs are in Appendix D.1. We also analyze transferability of f on arbitrary new distributions (x', y') information-theoretically in Appendix C.1, finding that a one-to-one encoder obeys the Infomax principle [7] better than collapsed embeddings on (x', y'). These observations suggest that a distribution over the embeddings that preserves strata distinctions and does not collapse classes is more desirable.

3.2. Modified Contrastive Loss L_{spread}

We introduce the loss L_{spread}, a weighted sum of two contrastive losses $L_{attract}$ and L_{repel}. $L_{attract}$ is a supervised contrastive loss, while L_{repel} encourages intra-class separation. For $\alpha \in [0,1]$,

$$L_{spread} = \alpha L_{attract} + (1-\alpha) L_{repel}. \tag{1}$$

For a given anchor x_i, define x_i^{aug} as an augmentation of the same point as x. Define the set of negative examples for i to be $N(i,B) = \{a \in B \backslash i : h(a) \neq h(i)\}$. Then,

$$\hat{L}_{attract}(f, x_i, B) = \frac{-1}{|P(i,B)|} \times \sum_{p \in P(i,B)} \log \frac{\exp(\sigma(x_i, x_p))}{\exp(\sigma(x_i, x_p)) + \sum_{a \in N(i,B)} \exp(\sigma(x_i, x_a))} \tag{2}$$

$$\hat{L}_{repel}(f, x_i, B) = -\log \frac{\exp(\sigma(x_i, x_i^{aug}))}{\sum_{p \in P(i,B)} \exp(\sigma(x_i, x_p))}. \tag{3}$$

$\hat{L}_{attract}$ is a variant of the SupCon loss, which encourages class separation in embedding space as suggested by Graf et al. [2]. $\hat{L}_{repel}$ is a class-conditional InfoNCE loss, where the positive distribution consists of augmentations and the negative distribution consists of i.i.d samples from the same class. It encourages points within a class to be spread apart, as suggested by the analysis of the InfoNCE loss by Wang and Isola [8].

Qualitative Evaluation

Figure 2 shows t-SNE plots for embeddings produced with L_{SC} versus L_{spread} on the CIFAR10 test set. L_{spread} produces embeddings that are more spread out than those produced by L_{SC} and avoids class collapse. As a result, images from different strata can be better differentiated in embedding space. For example, we show two dogs, one from a common stratum and one from a rare stratum (rare pose). The two dogs are much more distinguishable by distance in the L_{spread} embedding space, which suggests that it helps preserve distinctions between strata.

4. Geometry of Strata

We first discuss some existing theoretical tools for analyzing contrastive loss geometrically and their shortcomings with respect to understanding how strata are embedded. In Section 4.2, we propose a simple thought experiment about the distances between strata in embedding space when trained under a finite subsample of data to better understand our prior qualitative observations. Then, in Section 4.3, we discuss implications of representations that preserve strata distinctions, showing theoretically how they can yield better generalization error on both coarse-to-fine transfer and the original task and empirically how they allow for new downstream applications.

4.1. Existing Analysis

Previous works have studied the geometry of optimal embeddings under contrastive learning [2,8,9], but their techniques cannot analyze strata because strata information is not directly used in the loss function. These works use the *infinite encoder* assumption, where any distribution on $\mathbb{S}^{d-1}$ is realizable by the encoder f applied to the input data. This allows the minimization of the contrastive loss to be equivalent to an optimization problem over probability measures on the hypersphere. As a result, solving this new problem yields a distribution whose characterization is solely determined by information in the loss function (e.g., labels information [2,9]) and is decoupled from other information about the input data x and hence decoupled from strata.

More precisely, if we denote the measure of $x \in \mathcal{X}$ as $\mu_{\mathcal{X}}$, minimizing the contrastive loss over the mapping f is equal (at the population level) to minimizing over the pushforward measure $\mu_{\mathcal{X}} \circ f^{-1} : \mathbb{S}^{d-1} \to [0,1]$. The infinite encoder assumption allows us to relax the problem and instead consider optimizing over any $\mu \in \mathcal{M}(\mathbb{S}^{d-1})$ in the Borel set of probability measures on the hypersphere. Then, the optimal $\mu^{\star}$ learned is independent of the distribution of the input data $\mathcal{P}$ beyond what is in the relaxed objective function.

This approach using the infinite encoder assumption does not allow for analysis of strata. Strata are unknown at training time and thus cannot be incorporated explicitly into the loss function. Their geometries will not be reflected in the characterization of the optimal distribution obtained from previous theoretical tools. Therefore, we need additional reasoning for our empirical observations that strata distinctions are preserved in embedding space under L_{spread}.

4.2. Subsampling Strata

We propose a simple thought experiment based on *subsampling the dataset*—randomly sampling a fraction of the training data—to analyze strata. Consider the following: we subsample a fraction $t \in [0,1]$ of a training set of N points from $\mathcal{P}$. We use this subsampled dataset $\mathcal{D}_t$ to learn an encoder $\hat{f}_t$, and we study the average distance under $\hat{f}_t$ between two strata z and z' as t varies.

The average distance between z and z' is $\delta(\hat{f}_t, z, z') = \|\mathbb{E}_{x \sim \mathcal{P}_z}[\hat{f}_t(x)] - \mathbb{E}_{x \sim \mathcal{P}_{z'}}[\hat{f}_t(x)]\|_2$ and depends on whether z and z' are both in the subsampled dataset. We study when z and z' belong to the same class. We have three cases (with probabilities stated in Appendix C.2) based on strata frequency and t—when both, one, or neither of the strata appears in $\mathcal{D}_t$:

1. **Both strata appear in $\mathcal{D}_t$** The encoder $\hat{f}_t$ is trained on both z and z'. For large N, we can approximate this setting by considering $\hat{f}_t$ trained on infinite data from these strata. Points belonging to these strata will be defined in the optimal embedding distribution on the hypersphere, which can be characterized by prior theoretical approaches [2,8,9]. With L_{spread}, $\delta(\hat{f}_t, z, z')$ depends on α, which controls the extent of spread in the embedding geometry. With L_{SC}, points from the two strata would asymptotically map to one location on the hypersphere, and $\delta(\hat{f}_t, z, z')$ would converge to 0. This case occurs with probability increasing in $p(z), p(z')$, and t.
2. **One stratum but not the other appears in $\mathcal{D}_t$** Without loss of generality, suppose that points from z appear in $\mathcal{D}_t$ but no points from z' do. To understand $\delta(\hat{f}_t, z, z')$, we can consider how the end model $\hat{p}(y|\hat{f}_t(x))$ learned using the "source" distribution containing z performs on the "target" distribution of stratum z' since this downstream classifier is a function of distances in embedding space. Borrowing from the literature in domain adaptation, the difficulty of this out-of-distribution problem depends on both the divergence between source z and target z' distributions and the capacity of the overall model. The $\mathcal{H}\Delta\mathcal{H}$-divergence from Ben-David et al. [10,11], which is studied in lower bounds in Ben-David and Urner [12], and the discrepancy difference from Mansour et al. [13] capture both concepts. Moreover, the optimal geometries of L_{spread} and L_{SC} induce different end model capacities and prediction distributions, with data being more separable under L_{SC}, which can help explain why L_{spread} better preserves strata distances. This case occurs with probability increasing in $p(z)$ and decreasing in $p(z')$ and t.
3. **Neither strata appears in $\mathcal{D}_t$** The distance $\delta(\hat{f}_t, z, z')$ in this case is at most $2D_{\text{TV}}(\mathcal{P}_z, \mathcal{P}_{z'})$ (total variation distance) regardless of how the encoder is trained, although differences in transfer from models learned on $\mathcal{Z} \backslash z, z'$ to z versus z' can be further analyzed. This case occurs with probability decreasing in $p(z), p(z')$, and t.

We make two observations from these cases. First, if z and z' are both common strata, then as t increases, the distance between them depends on the optimal asymptotic distribution. Therefore, if we set $\alpha = 1$ in L_{spread}, these common strata will collapse. Second, if z is a common strata and z' is uncommon, the second case occurs frequently over randomly sampled $\mathcal{D}_t$, and thus the strata are separated based on the difficulty of the respective out-of-distribution problem. We thus arrive at the following insight from our thought experiment:

Common strata are more tightly clustered together, while rarer and more semantically distinct strata are far away from them.

Figure 3 demonstrates this insight. It shows a t-SNE visualization of embeddings from training on CIFAR100 with coarse superclass labels, and with artifically imbalanced subclasses. We show points from the largest subclasses in dark blue and points from the smallest subclasses in light blue. Points from the largest subclasses (dark blue) cluster tightly, whereas points from small subclasses (light blue) are scattered throughout the embedding space.

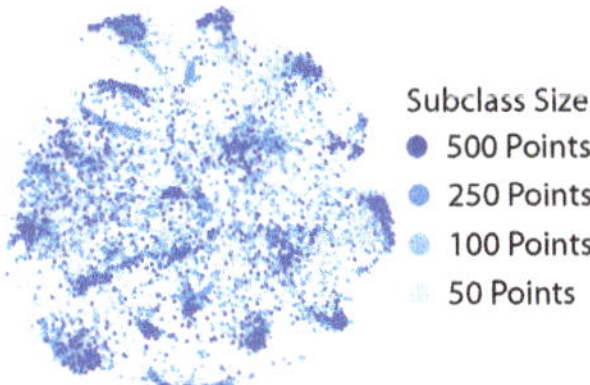

Figure 3. Points from large subclasses cluster tightly; points from small subclasses scatter (CIFAR100-Coarse, unbalanced subclasses).

4.3. *Implications*

We discuss theoretical and practical implications of our subsampling argument. First, we show that on both the coarse-to-fine transfer task (x, z) and the original task (x, y), embeddings that preserve strata yield better generalization error. Second, we discuss practical implications arising from our subsampling argument that enable new applications.

4.3.1. Theoretical Implications

Consider $\hat{f}_1$, the encoder trained on $\mathcal{D}$ with all N points using L_{spread}, and suppose a mean classifier is used for the end model, e.g., $W_y = \mathbb{E}_{x|y}\left[\hat{f}_1(x)\right]$ and $W_z = \mathbb{E}_{x|z}\left[\hat{f}_1(x)\right]$. On coarse-to-fine transfer, generalization error depends on how far each stratum center is from the others.

Lemma 1. *There exists $\lambda_z > 0$ such that the generalization error on the coarse-to-fine transfer task is at most*

$$\mathcal{L}(x, z, \hat{f}_1) \leq \mathbb{E}_z\left[\log\left(\sum_{z' \in \mathcal{Z}} \exp\left(-\lambda_z\left(\frac{1}{2}\delta(\hat{f}_1, z, z')^2 - 1\right)\right)\right)\right] - 1, \tag{4}$$

where $\delta(\hat{f}_1, z, z')$ is the average distance between strata z and z' defined in Section 4.2.

The larger the distances between strata, the smaller the upper bound on generalization error. We now show that a similar result holds on the original task (x, y), but there is an additional term that penalizes points from the same class being too far apart.

Lemma 2. *There exists $\lambda_y > 0$ such that the generalization error on the original task is at most*

$$\mathcal{L}(x, y, \hat{f}_1) \leq \mathbb{E}_z\left[\mathbb{E}_{z'|S(z)}\left[\frac{1}{2}\delta(\hat{f}_1, z, z')^2 - 1\right]\right. \tag{5}$$

$$\left. + \log\left(\sum_{y \in \mathcal{Y}} \exp\left(\mathbb{E}_{z'|y}\left[-\lambda_y\left(\frac{1}{2}\delta(\hat{f}_1, z, z')^2 - 1\right)\right]\right)\right)\right]. \tag{6}$$

This result suggests that maximizing distances between strata of different classes is desirable, but less so for distances between strata of the same class as suggested by the first term in the expression. Both results illustrate that separating strata to some extent in the embedding space results in better bounds on generalization error. In Appendix C.3, we provide proofs of these results and derive values of the generalization error for these two tasks under class collapse for comparison.

4.3.2. Practical Implications

Our discussion in Section 4.2 suggests that training with L_{spread} better distinguishes strata in embedding space. As a result, we can use differences between strata of different sizes for downstream applications. For example, unsupervised clustering can help recover pseudolabels for unlabeled, rare strata. These pseudolabels can be used as inputs to worst-group robustness algorithms, or used to detect noisy labels, which appear to be rare strata during training (see Section 5.3 for examples). We can also train over subsampled datasets to heuristically distinguish points that come from common strata from points that come from rare strata. We can then downsample points from common strata to construct minimal coresets (see Section 5.4 for examples).

5. Experiments

This section evaluates L_{spread} on embedding quality and model quality:

- First, in Section 5.2, we use coarse-to-fine transfer learning to evaluate how well the embeddings maintain strata information. We find that L_{spread} achieves lift across four datasets.
- In Section 5.3, we evaluate how well L_{spread} can detect rare strata in an unsupervised setting. We first use L_{spread} to detect rare strata to improve worst-group robustness by up to 2.5 points. We then use rare strata detection to correct noisy labels, recovering 75% performance under 20% noise.
- In Section 5.4, we evaluate how well L_{spread} can distinguish points from large strata versus points from small strata. We downsample points from large strata to construct minimal coresets on CIFAR10, outperforming prior work by 1.0 points at 30% labeled data.
- Finally, in Section 5.5, we show that training with L_{spread} improves model quality, validating our theoretical claims that preventing class collapse can improve generalization error. We find that L_{spread} improves performance in 7 out of 9 cases.

5.1. Datasets and Models

Tabel 1 lists all the datasets we use in our evaluation. CIFAR10, CIFAR100, and MNIST are the standard computer vision datasets. We also use coarse versions of each, wherein classes are combined to create coarse superclasses (animals/vehicles for CIFAR10, standard superclasses for CIFAR100, and <5, ≥ 5 for MNIST). In CIFAR100-Coarse-U, some subclasses have been artificially imbalanced. Waterbirds, ISIC and CelebA are image datasets with documented hidden strata [5,14–16]. We use a ViT model [17] (4×4, 7 layers) for CIFAR and MNIST and a ResNet50 for the rest. For the ViT models, we jointly optimize the contrastive loss with a cross entropy loss head. For the ResNets, we train the contrastive loss on its own and use linear probing on the final layer. More details in Appendix E.

Table 1. Summary of the datasets we use for evaluation.

Dataset	Notes
CIFAR10	Standard computer vision dataset
CIFAR10-Coarse	CIFAR10 with animal/vehicle coarse labels
CIFAR100	Standard computer vision dataset
CIFAR100-Coarse	CIFAR100 with standard coarse labels
CIFAR100-Coarse-U	CIFAR100 with standard coarse labels, but with some fine classes sub-sampled
MNIST	Standard computer vision dataset
MNIST-Coarse	MNIST with <5 and ≥5 coarse labels
Waterbirds	Robustness dataset mixing up images of birds and their backgrounds [14]
ISIC	Images of skin lesions [15]
CelebA	Images of celebrity faces [16]

5.2. Coarse-to-Fine Transfer Learning

In this section, we use coarse-to-fine transfer learning to evaluate how well L_{spread} retains strata information in the embedding space. We train on coarse superclass labels, freeze the weights, and then use transfer learning to train a linear layer with subclass labels. We use this supervised strata recovery setting to isolate how well the embeddings can recover strata in the optimal setting. For baselines, we compare against training with L_{SC} and the SimCLR loss L_{SS}.

Table 2 reports the results. We find that L_{spread} produces better embeddings for coarse-to-fine transfer learning than L_{SC} and L_{SS}. Lift over L_{SC} varies from 0.2 points on MNIST (16.7% error reduction), to 23.6 points of lift on CIFAR10. L_{spread} also produces better embeddings than L_{SS}, since L_{SS} does not encode superclass labels in the embedding space.

Table 2. Performance of coarse-to-fine transfer on various datasets compared against contrastive baselines. In these tasks, we first train a model on coarse task labels, then freeze the representation and train a model on fine-grained subclass labels. L_{spread} produces embeddings that transfer better across all datasets. Best in bold.

	Coarse-to-Fine Transfer		
Dataset	L_{SS}	L_{SC}	L_{spread}
CIFAR10-Coarse	71.7	52.5	**76.1**
CIFAR100-Coarse	62.0	62.4	**63.9**
CIFAR100-Coarse-U	61.9	59.5	**62.4**
MNIST-Coarse	97.1	98.8	**99.0**

5.3. Robustness Against Worst-Group Accuracy and Noise

In this section, we use robustness to measure how well L_{spread} can recover strata in an unsupervised setting. We use clustering to detect rare strata as an input to worst-group robustness algorithms, and we use a geometric heuristic over embeddings to correct noisy labels.

To evaluate worst-group accuracy, we follow the experimental setup and datasets from Sohoni et al. [5]. We first train a model with class labels. We then cluster the embeddings to produce pseudolabels for hidden strata, which we use as input for a Group-DRO algorithm to optimize worst-group robustness [14]. We use both L_{SC} and cross entropy loss [5] for training the first stage as baselines.

To evaluate robustness against noise, we introduce noisy labels to the contrastive loss head on CIFAR10. We detect noisy labels with a simple geometric heuristic: points with incorrect labels appear to be small strata, so they should be far away from other points of the same class. We then correct noisy points by assigning the label of the nearest cluster in the batch. More details can be found in Appendix E.

Table 3 shows the performance of unsupervised strata recovery and downstream worst-group robustness. We can see that L_{spread} outperforms both L_{SC} and Sohoni et al. [5] on strata recovery. This translates to better worst-group robustness on Waterbirds and CelebA.

Figure 4 (left) shows the effect of noisy labels on performance. When noisy labels are uncorrected (purple), performance drops by up to 10 points at 50% noise. Applying our geometric heuristic (red) can recover 4.8 points at 50% noise, even without using L_{spread}. However, L_{spread} recovers an additional 0.9 points at 50% noise, and an additional 1.6 points at 20% noise (blue). In total, L_{spread} recovers 75% performance at 20% noise, whereas L_{SC} only recovers 45% performance.

Table 3. Unsupervised strata recovery performance (top, F1), and worst-group performance (AUROC for ISIC, Acc for others) using recovered strata. Best in bold.

Dataset	Sohoni et al. [5]	L_{SC}	L_{spread}
	Sub-Group Recovery		
Waterbirds	56.3	47.2	**59.0**
ISIC	74.0	92.5	**93.8**
CelebA	24.2	19.4	**24.8**
	Worst-Group Robustness		
Waterbirds	88.4	86.5	**89.0**
ISIC	92.0	**93.3**	92.6
CelebA	55.0	66.1	**67.8**

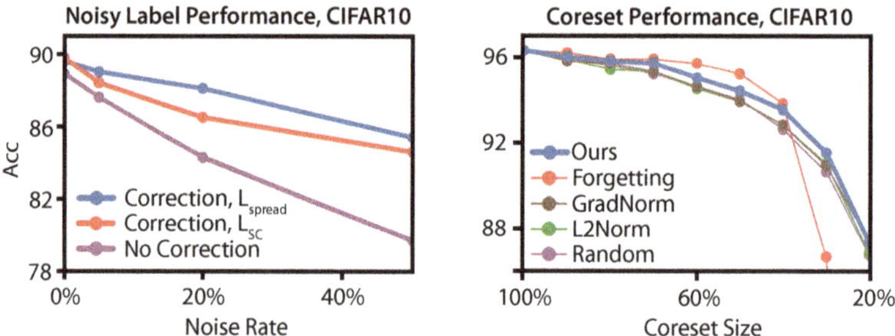

Figure 4. (**Left**) Performance of models under various amounts of label noise for the contrastive loss head. (**Right**) Performance of a ResNet18 trained with coresets of various sizes. Our coreset algorithm is competitive with the state-of-the-art in the large coreset regime (from 40–90% coresets), but maintains performance for small coresets (smaller than 40%). At the 10% coreset, our algorithm outperforms [18] by 32 points and matches random sampling.

5.4. Minimal Coreset Construction

Now we evaluate how well training on fractional samples of the dataset with L_{spread} can distinguish points from large versus small strata by constructing minimal coresets for CIFAR10. We train a ResNet18 on CIFAR10, following Toneva et al. [18], and compare against baselines from Toneva et al. [18] (Forgetting) and Paul et al. [19] (GradNorm, L2Norm). For our coresets, we train with L_{spread} on subsamples of the dataset and record how often points are correctly classified at the end of each run. We bucket points in the training set by how often the point is correctly classified. We then iteratively remove points from the largest bucket in each class. Our strategy removes easy examples first from the largest coresets, but maintains a set of easy examples in the smallest coresets.

Figure 4 (right) shows the results at various coreset sizes. For large coresets, our algorithm outperforms both methods from Paul et al. [19] and is competitive with Toneva et al. [18]. For small coresets, our method outperforms the baselines, providing up to 5.2 points of lift over Toneva et al. [18] at 30% labeled data. Our analysis helps explain this gap; removing too many easy examples hurts performance, since then the easy examples become rare and hard to classify.

5.5. Model Quality

Finally, we confirm that L_{spread} produces higher-quality models and achieves better sample complexity than both L_{SC} and the SimCLR loss L_{SS} from [20]. Table 4 reports the performance of models across all our datasets. We find that L_{spread} achieves better overall performance compared to models trained with L_{SC} and L_{SS} in 7 out of 9 tasks, and matches performance in 1 task. We find up to 4.0 points of lift over L_{SC} (Waterbirds), and up to 2.2 points of lift (AUROC) over L_{SS} (ISIC). In Appendix F, we additionally evaluate the sample complexity of contrastive losses by training on partial subsamples of CIFAR10. L_{spread} outperforms L_{SC} and L_{SS} throughout.

Table 4. End model performance training with L_{spread} on various datasets compared against contrastive baselines. All metrics are accuracy except for ISIC (AUROC). L_{spread} produces the best performance in 7 out of 9 cases, and matches the best performance in 1 case. Best in bold.

	End Model Perf.		
Dataset	L_{SS}	L_{SC}	L_{spread}
CIFAR10	89.7	90.9	**91.5**
CIFAR10-Coarse	97.7	96.5	**98.1**
CIFAR100	68.0	67.5	**69.1**
CIFAR100-Coarse	76.9	77.2	**78.3**
CIFAR100-Coarse-U	72.1	71.6	**72.4**
MNIST	99.1	**99.3**	99.2
MNIST-Coarse	99.1	**99.4**	**99.4**
Waterbirds	77.8	73.9	**77.9**
ISIC	87.8	88.7	**90.0**

6. Related Work and Discussion

From work in **contrastive learning**, we take inspiration from [21], who use a latent classes view to study self-supervised contrastive learning. Similarly, [22] considers how minimizing the InfoNCE loss recovers a latent data generating model. We initially started from a debiasing angle to study the effects of noise in supervised contrastive learning inspired by [23], but moved to our current strata-based view of noise instead. Recent work has also analyzed contrastive learning from the information-theoretic perspective [24–26], but does not fully explain practical behavior [27], so we focus on the geometric perspective in this paper because of the downstream applications. On the geometric side, we are inspired by the theoretical tools from [8] and [2], who study representations on the hypersphere along with [9].

Our work builds on the recent wave of empirical interest in contrastive learning [20,28–31] and supervised contrastive learning [1]. There has also been empirical work analyzing the transfer performance of contrastive representations and the role of intra-class variability in transfer learning. [32] find that combining supervised and self-supervised contrastive loss improves transfer learning performance, and they hypothesize that this is due to both inter-class separation and intra-class variability. [33] find that combining cross entropy and self-supervised contrastive loss improves coarse-to-fine transfer, also motivated by preserving intra-class variability.

We derive L_{spread} from similar motivations to losses proposed in these works, and we futher theoretically study why class collapse can hurt downstream performance. In particular, we study why preserving distinctions of strata in embedding space may be important, with theoretical results corroborating their empirical studies. We further propose a new thought experiment for why a combined loss function may lead to better separation of strata.

Our treatment of **strata** is strongly inspired by [5,6], who document empirical consequences of hidden strata. We are inspired by empirical work that has demonstrated that detecting subclasses can be important for performance [4,34] and robustness [14,35,36].

Each of our downstream **applications** is a field in itself, and we take inspiration from recent work from each. Our noise heuristic is similar to the ELR [37] and takes inspiration from a various work using contrastive learning to correct noisy labels and for semi-supervised learning [38–40]. Our coreset algorithm is inspired by recent work in coresets for modern deep networks [19,41,42], and takes inspiration from [18] in particular.

7. Conclusions

We propose a new supervised contrastive loss function to prevent class collapse and produce higher-quality embeddings. We discuss how our loss function better maintains strata distinctions in embedding space and explore several downstream applications. Future directions include encoding label hierarchies and other forms of knowledge in contrastive loss functions and extending our work to more modalities, models, and applications. We hope that our work inspires further work in more fine-grained supervised contrastive loss functions and new theoretical approaches for reasoning about generalization and strata.

Author Contributions: Conceptualization, D.Y.F. and M.F.C.; methodology, D.Y.F. and M.F.C.; software, D.Y.F.; validation, D.Y.F. and M.Z.; formal analysis, M.F.C.; investigation, D.Y.F., M.F.C. and M.Z.; resources, D.Y.F. and M.F.C.; data curation, D.Y.F.; writing—original draft preparation, D.Y.F., M.F.C. and M.Z.; writing—review and editing, D.Y.F., M.F.C. and M.Z.; visualization, D.Y.F.; supervision, K.F. and C.R.; project administration, D.Y.F. and M.F.C.; funding acquisition, D.Y.F. and M.F.C. All authors have read and agreed to the published version of the manuscript.

Funding: This research was funded by NIH under No. U54EB020405 (Mobilize), NSF under Nos. CCF1763315 (Beyond Sparsity), CCF1563078 (Volume to Velocity), and 1937301 (RTML); ONR under No. N000141712266 (Unifying Weak Supervision); ONR N00014-20-1-2480: Understanding and Applying Non-Euclidean Geometry in Machine Learning; N000142012275 (NEPTUNE); the Moore Foundation, NXP, Xilinx, LETI-CEA, Intel, IBM, Microsoft, NEC, Toshiba, TSMC, ARM, Hitachi, BASF, Accenture, Ericsson, Qualcomm, Analog Devices, the Okawa Foundation, American Family Insurance, Google Cloud, Salesforce, Total, the HAI-GCP Cloud Credits for Research program, the Stanford Data Science Initiative (SDSI), Department of Defense (DoD) through the National Defense Science and Engineering Graduate Fellowship (NDSEG) Program, and members of the Stanford DAWN project: Facebook, Google, and VMWare. The Mobilize Center is a Biomedical Technology Resource Center, funded by the NIH National Institute of Biomedical Imaging and Bioengineering through Grant P41EB027060. The U.S. Government is authorized to reproduce and distribute reprints for Governmental purposes notwithstanding any copyright notation thereon. Any opinions, findings, and conclusions or recommendations expressed in this material are those of the authors and do not necessarily reflect the views, policies, or endorsements, either expressed or implied, of NIH, ONR, or the U.S. Government.

Institutional Review Board Statement: Not Applicable.

Informed Consent Statement: Not Applicable.

Data Availability Statement: Datasets used in this paper are publicly available and described in Appendix E.

Acknowledgments: We thank Nimit Sohoni for helping with coreset and robustness experiments, and we thank Beidi Chen and Tri Dao for their helpful comments.

Conflicts of Interest: The authors declare no conflict of interest. The funders had no role in the design of the study; in the collection, analyses, or interpretation of data; in the writing of the manuscript, or in the decision to publish the results.

We provide a glossary in Appendix A. Then we provide definitions of terms in Appendix B. We discuss additional theoretical results in Appendix C. We provide proofs in Appendix D. We discuss additional experimental details in Appendix E. Finally, we provide additional experimental results in Appendix F.

Appendix A. Glossary

The glossary is given in Table A1 below.

Table A1. Glossary of variables and symbols used in this paper.

Symbol	Used for
L_{SC}	SupCon (see Section 2.2), a supervised contrastive loss introduced by [1].
L_{spread}	Our modified loss function defined in Section 3.2.
x	Input data $x \in \mathcal{X}$.
y	Class label $y \in \mathcal{Y} = \{1, \dots, K\}$.
$\mathcal{D}$	Dataset of N points $\{(x_i, y_i)\}_{i=1}^N$ drawn i.i.d. from $\mathcal{P}$.
$h(x)$	The class that x belongs to, i.e., $h(x)$ is a label drawm from $p(y\|x)$. This label information is used as input in the supervised contrastive loss.
$\hat{p}(y\|x)$	The end model's predicted distribution over y given x.
z	A stratum is a latent variable $z \in \mathcal{Z} = \{1, \dots, C\}$ that further categorizes data beyond labels.
S_k	The set of all strata corresponding to label k (deterministic).
$S(c)$	The label corresponding to strata c (deterministic).
$\mathcal{P}_z$	The distribution of input data belonging to stratum z, i.e., $x \sim p(\cdot\|z)$.
m	The number of strata per class.
d	Dimension of the embedding space.
f	The encoder $f : \mathcal{X} \to \mathbb{R}^d$ maps input data to an embedding space and is learned by minimizing the contrastive loss function.
$\mathbb{S}^{d-1}$	The unit hypersphere, formally $\{v \in \mathbb{R}^d : \\|v\\|_2 = 1\}$.
τ	Temperature hyperparameter in contrastive loss function.
$\sigma(x, x')$	Notation for $\frac{f(x)^\top f(x')}{\tau}$.
$\mathcal{B}$	Set of batches of labeled data on $\mathcal{D}$.
$P(i, B)$	Points in B with the same label as x_i, formally $\{p \in B \backslash i : h(p) = h(i)\}$.
$\{v_i\}_{i=1}^K$	A regular simplex inscribed in the hypersphere (see Definition A1).
W	The weight matrix that parametrizes the downstream linear classifier (end model) learned on $f(x)$.
$\hat{\mathcal{L}}(W, \mathcal{D})$	The empirical cross entropy loss used to learn W over dataset $\mathcal{D}$ (see (A1)).
$\mathcal{L}(x, y, f)$	The generalization error of the end model of predicting output y on x using encoder f (see (A2) and (A3)).
$L_{attract}$	A variant on SupCon that is used in L_{spread} that pushes points of a class together (see (2)).
L_{repel}	A class-conditional InfoNCE loss that is used in L_{spread} to pull apart points within a class (see (3)).
α	Hyperparameter $\alpha \in [0, 1]$ controls how to balance $L_{attract}$ and L_{repel}.
x^{aug}	An augmentation of data point x.
$N(i, B)$	Points in B with a label different from that of x_i, formally $\{a \in B \backslash i : h(a) \neq h(i)\}$.
t	Fraction of training data $t \in [0, 1]$ that is varied in our thought experiment.
$\mathcal{D}_t$	Randomly sampled dataset from $\mathcal{P}$ with size equal to $t \cdot N$ fraction of $\mathcal{D}$.
$\hat{f}_t$	Encoder trained on sampled dataset $\mathcal{D}_t$.
$\delta(\hat{f}_t, z, z')$	The distance between centers of strata z and z' under encoder $\hat{f}_t$, namely $\delta(\hat{f}_t, z, z') = \\|\mathbb{E}_{x \sim \mathcal{P}_z}[\hat{f}_t(x)] - \mathbb{E}_{x \sim \mathcal{P}_{z'}}[\hat{f}_t(x)]\\|_2$.

Appendix B. Definitions

We restate definitions used in our proofs.

Definition A1 (Regular Simplex). *The points $\{v_i\}_{i=1}^K$ form a regular simplex inscribed in the hypersphere if*

1. $\sum_{i=1}^K v_i = 0$

2. $\|v_i\| = 1$ *for all i*
3. $\exists c_K \leq 1$ s.t. $v_i^\top v_j = c_K$ *for* $i \neq j$

Definition A2 (Downstream model). *Once an encoder $f(x)$ is learned, the downstream model consists of a linear classifier trained using the cross-entropy loss:*

$$\hat{\mathcal{L}}(W, \mathcal{D}) = \sum_{x_i \in \mathcal{D}} -\log \frac{\exp(f(x_i)^\top W_{h(x_i)})}{\sum_{j=1}^K \exp(f(x_i)^\top W_j)}. \tag{A1}$$

Define $\hat{W} := \text{argmin}_{\|W\|^2 \leq 1} \hat{\mathcal{L}}(W, \mathcal{D})$. Then, the end model's outputs are the probabilities

$$\hat{p}(y|x) = \hat{p}(y|f(x)) = \frac{\exp(f(x)^\top \hat{W}_y)}{\sum_{j=1}^K, \exp(f(x)^\top \hat{W}_j)}, \tag{A2}$$

and the generalization error is

$$\mathcal{L}(x, y, f) = \mathbb{E}_{x,y}[-\log \hat{p}(y|f(x))]. \tag{A3}$$

Appendix C. Additional Theoretical Results

Appendix C.1. Transfer Learning on (x', y')

We now show an additional transfer learning result on new tasks (x', y'). Formally, recall that we learn the encoder f on $(x, y) \sim \mathcal{P}$. We wish to use it on a new task with target distribution $(x', y') \sim \mathcal{P}'$. We find that an injective encoder $f(x)$ is more appropriate to be used on new distributions than collapsed embeddings based on the Infomax principle [7].

Observation 3. *Define $f_c(y)$ as the mapping to collapsed embeddings and $f_{1-1}(x)$ as an injective mapping, both learned on $\mathcal{P}$. Construct a new variable $\tilde{y}$ with joint distribution $(x', \tilde{y}) \sim p(y|x) \cdot p'(x')$ and suppose that $\tilde{y} \perp\!\!\!\perp y'|x'$. Then, by the data processing inequality, it holds that $I(\tilde{y}, y') \leq I(x', y')$ where $I(\cdot, \cdot)$ is the mutual information between two random variables. We apply f_c to $\tilde{y}$ and f_{1-1} to x' to get that*

$$I(f_c(\tilde{y}), y') \leq I(f_{1-1}(x'), y').$$

Therefore, f_{1-1} obeys the Infomax principle [7] better on $\mathcal{P}'$ than f_c. Via Fano's inequality, this statement implies that the Bayes risk for learning y' from x' is lower using f_{1-1} than f_c.

Appendix C.2. Probabilities of Strata z, z' Appearing in Subsampled Dataset

As discussed in Section 4.2, the distance between strata z and z' in embedding space depends on if these strata appear in the subsampled dataset $\mathcal{D}_t$ that the encoder was trained on. We define the exact probabilities of the three cases presented. Let $\Pr(z, z' \in \mathcal{D}_t)$ be the probability that both strata are seen, $\Pr(z \in \mathcal{D}_t, z' \notin \mathcal{D}_t)$ be the probability that only z is seen, and $\Pr(z, z' \notin \mathcal{D}_t$ be the probability that neither are seen.

First, the probability of neither strata appearing in $\mathcal{D}_t$ is easy to compute. In particular, we have that $\Pr(z, z' \notin \mathcal{D}_t) = (1 - p(z) - p(z'))^{tN}$. This quantity decreases in $p(z)$ and $p(z')$, confirming that it is less likely for two common strata to not appear in $\mathcal{D}_t$.

Second, the probability of z being in $\mathcal{D}_t$ and z' not being in $\mathcal{D}_t$ can be expressed as $\Pr(z \in \mathcal{D}_t | z' \notin \mathcal{D}_t) \cdot \Pr(z' \notin \mathcal{D}_t)$. $\Pr(z' \notin \mathcal{D}_t)$ is equal to $(1 - p(z'))^{tN}$, and $\Pr(z \in \mathcal{D}_t | z' \notin \mathcal{D}_t) = 1 - \Pr(z \notin \mathcal{D}_t | z' \notin \mathcal{D}_t) = 1 - (1 - p(z | z \in \mathcal{Z} \backslash z'))^{tN}$. Finally, note that $p(z | z \in \mathcal{Z} \backslash z') = \frac{p(z)}{1 - p(z')}$. Putting this together, we get that $\Pr(z \in \mathcal{D}_t, z' \notin \mathcal{D}_t) = (1 - p(z'))^{tN} - (1 - p(z') - p(z))^{tN}$, and we can similarly construct $\Pr(z' \in \mathcal{D}_t, z \notin \mathcal{D}_t)$.

This quantity depends on the difference between $p(z)$ and $p(z')$, so this case is common when one stratum is common and one is rare.

Lastly, the probability of both z and z′ being in $\mathcal{D}_t$ is thus $\Pr(z, z' \in \mathcal{D}_t) = 1 - \Pr(z, z' \notin \mathcal{D}_t) - \Pr(z' \in \mathcal{D}_t, z \notin \mathcal{D}_t) - \Pr(z \in \mathcal{D}_t, z' \notin \mathcal{D}_t) = 1 + (1 - p(z') - p(z))^{tN} - (1 - p(z'))^{tN} - (1 - p(z))^{tN}$. This quantity increases in $p(z)$ and $p(z')$.

Appendix C.3. Performance of Collapsed Embeddings on Coarse-to-Fine Transfer and Original Task

Lemma A1. *Denote f_c to be the encoder that collapses embeddings such that $f_c(x) = v_y$ for any $(x, y) \sim \mathcal{P}$. Then, the generalization error on the coarse-to-fine transfer task using f_c and a linear classifier learned using cross entropy loss is at least*

$$\mathcal{L}(x, z, f_c) \geq \log(m \exp(1) + (C - m) \exp(c_K) - 1,$$

where c_K is the dot product of any two different class-collapsed embeddings. The generalization error on the original task under the same setup is at least

$$\mathcal{L}(x, y, f_c) \geq \log(\exp(1) + (K - 1) \exp(c_K)) - 1.$$

Proof. We first bound generalization error on the coarse-to-fine transfer task. For collapsed embeddings, $f(x) = v_i$ when $h(x) = i$, where $h(x)$ is information available at training time that follows the distribution $p(y|x)$. We thus denote the embedding $f(x)$ as $v_{h(x)}$. Therefore, we write the generalization error with an expectation over $h(x)$ and factorize the expectation according to our generative model.

$$\begin{aligned}
\mathbb{E}_{x,z,h(x)}[-\log \hat{p}(z|f(x))] &= -\sum_{z=1}^{C} \sum_{h(x)=1}^{K} \int p(x, z, h(x)) \log \hat{p}(z|h(x)) dx \\
&= -\sum_{z=1}^{C} \sum_{h(x)=1}^{K} \int p(z) p(x|z) p(h(x)|x) \log \hat{p}(z|h(x)) dx \\
&= -\sum_{z=1}^{C} \sum_{h(x)=1}^{K} \int p(z) p(x|z) p(h(x)|x) \log \frac{\exp(f_{h(x)}^\top W_z)}{\sum_{i=1}^{C} \exp(f_{h(x)}^\top W_i)} dx \\
&= \sum_{z=1}^{C} p(z) \mathbb{E}_{x \sim \mathcal{P}_z} \left[\sum_{y=1}^{K} p(y|x) \left(-v_y^\top W_z + \log \sum_{i=1}^{C} \exp(v_y^\top W_i) \right) \right].
\end{aligned}$$

Furthermore, since the W learned over collapsed embeddings satisfies $W_z = v_y$ for $S(z) = y$, we have that $\log \sum_{i=1}^{C} \exp(v_y^\top W_i) = m \exp(1) + (C - m) \exp(c_K)$ for any y, and our expected generalization error is

$$\begin{aligned}
&\sum_{z=1}^{C} p(z) \mathbb{E}_{x \sim \mathcal{P}_z}[-p(y = S(z)|x) - p(y \neq S(z)|x)\delta + \log(m \exp(1) + (C - m) \exp(c_K))] \\
&= \log(m \exp(1) + (C - m) \exp(c_K)) - c_K - (1 - c_K) \sum_{z=1}^{C} p(z) \mathbb{E}_{x \sim \mathcal{P}_z}[p(y = S(z)|x)].
\end{aligned}$$

This tells us that the generalization error is at most $\log(m \exp(1) + (C - m) \exp(c_K)) - c_K$ and at least $\log(m \exp(1) + (C - m) \exp(c_K)) - 1$.

For the original task, we can apply this same approach to the case where $m = 1, C = K$ to get that the average generalization error is

$$\begin{aligned}
\mathbb{E}_{h(x)} \left[\mathcal{L}(x, y, \hat{f}_1) \right] =& \log(\exp(1) + (K - 1) \exp(c_K)) \\
&- c_K - (1 - c_K) \sum_{z=1}^{C} p(z) \mathbb{E}_{x \sim \mathcal{P}_z}[p(y = S(z)|x)].
\end{aligned}$$

This is at least $\log(\exp(1) + (K-1)\exp(c_K)) - 1$ and at most $\log(\exp(1) + (K-1)\exp(c_K)) - c_K$. □

Appendix D. Proofs

Appendix D.1. Proofs for Theoretical Motivation

We provide proofs for Section 3.1. First, we characterize the optimal linear classifier (for both the coarse-to-fine transfer task and the original task) learned on the collapsed embeddings. Note that this result appears similar to Corollary 1 of [2], but their result minimizes the cross entropy loss over both the encoder and downstream weights (i.e., in a classical supervised setting where only cross entropy is used in training).

Lemma A2 (Downstream linear classifier for coarse-to-fine task). *Suppose the dataset $\mathcal{D}_z$ is class-balanced across z, and the embeddings satisfy $f(x) = v_i$ if $h(x) = i$ where $\{v_i\}_{i=1}^K$ form the regular simplex. Then the optimal weight matrix $W^\star \in \mathbb{R}^{C\times d}$ that minimizes $\hat{\mathcal{L}}(W, \mathcal{D}_z)$ satisfies $W_z^\star = v_y$ for $y = S(z)$.*

Proof. Formally, the convex optimization problem we are solving is

$$\text{minimize } -\sum_{y=1}^{K}\sum_{z\in S_y}\log\frac{\exp(v_y^\top W_z)}{\sum_{j=1}^{C}\exp(v_y^\top W_j)} \tag{A4}$$

$$\text{s.t. } \|W_z\|_2^2 \le 1 \;\forall z \in \mathcal{Z} \tag{A5}$$

The Lagrangian of this optimization problem is

$$\sum_{y=1}^{K}\sum_{z\in S_y} -v_y^\top W_z + m\sum_{y=1}^{K}\log\left(\sum_{j=1}^{C}\exp(v_y^\top W_j)\right) + \sum_{i=1}^{C}\lambda_i(\|W_i\|_2^2 - 1),$$

and the stationarity condition w.r.t. W_z is

$$-v_{S(z)} + m\sum_{y=1}^{K}\frac{v_y\exp(v_y^\top W_z)}{\sum_{j=1}^{C}\exp(v_y^\top W_j)} + 2\lambda_z W_z = 0. \tag{A6}$$

Substituting $W_z = v_{S(z)}$, we get $-v_{S(z)} + m\sum_{y=1}^{K}\frac{v_y\exp(v_y^\top v_{S(z)})}{\sum_{j=1}^{C}\exp(v_y^\top v_{S(j)})} + 2\lambda_z v_{S(z)} = 0$. Using the fact that $v_i^\top v_j = \delta$ for all $i \neq j$, this equals $-v_{S(z)} + m\cdot\frac{v_{S(z)}\exp(1)+\exp(\delta)\sum_{y\neq S(z)}v_y}{m\exp(1)+(C-m)\exp(\delta)} + 2\lambda_z v_{S(z)} = 0$. Next, recall that $\sum_{i=1}^{K} v_i = 0$. Then, $\lambda_z = \frac{1}{2}\left(1 - m\cdot\frac{\exp(1)-\exp(\delta)}{m\exp(1)+(C-m)\exp(\delta)}\right) \ge 0$, satisfying the dual constraint. We can further verify complementary slackness and primal feasibility, since $\|W_z^\star\|_2^2 = 1$, to confirm that an optimal weight matrix satisfies $W_z^\star = v_y$ for $y = S(z)$. □

Corollary A1. *When we apply the above proof to the case when $m = 1$, we recover that the optimal weight matrix $W^\star \in \mathbb{R}^{K\times d}$ that minimizes $\hat{\mathcal{L}}(W, \mathcal{D})$ for the original task on $(x, y) \sim \mathcal{P}$ satisfies $W_y^\star = v_y$ for all $y \in \mathcal{Y}$.*

We now prove Observation 1 and 2. Then, we present an additional result on transfer learning on collapsed embeddings to general tasks of the form $(x', y') \sim \mathcal{P}'$.

Proof of Observation 1. We write out the generalization error for the downstream task, $\mathcal{L}(x,z,f) = \mathbb{E}_{x,z}[-\log \hat{p}(z|x)]$ using our conditions that $p(y = h(x)|x) = 1$ and $p(z|x) = \frac{1}{m}$.

$$\begin{aligned}
\mathcal{L}(x,z,f) &= -\int p(x) \sum_{z=1}^{C} p(z|x) \log \hat{p}(z|f(x))dx \\
&= -\int p(x) \sum_{z=1}^{C} p(z|x) \log \frac{\exp(f(x)^\top W_z)}{\sum_{i=1}^{C} \exp(f(x)^\top W_i)} dx \\
&= -\sum_{y=1}^{K} \int_{x:h(x)=y} p(x) \cdot \frac{1}{m} \sum_{z\in S_y} \log \frac{\exp(f(x)^\top W_z)}{\sum_{i=1}^{C} \exp(f(x)^\top W_i)}.
\end{aligned}$$

To minimize this, $f(x)$ should be the same across all x where $h(x)$ is the same value, since $p(z|x)$ does not change across fixed $h(x)$ and thus varying $f(x)$ will not further decrease the value of this expression. Therefore, we rewrite $f(x)$ as $f_{h(x)}$. Using the fact that y is class balanced, our loss is now

$$\begin{aligned}
\mathcal{L}(x,y,z) &= -\frac{1}{m} \sum_{y=1}^{K} \sum_{z\in S_y} \int_{x:h(x)=y} p(x) \log \frac{\exp(f_{h(x)}^\top W_z)}{\sum_{i=1}^{C} \exp(f_{h(x)}^\top W_z)} dx \\
&= -\frac{1}{C} \sum_{y=1}^{K} \sum_{z\in S_y} \log \frac{\exp(f_y^\top W_z)}{\sum_{i=1}^{C} \exp(f_y^\top W_i)}.
\end{aligned}$$

We claim that $f_y = v_y$ and $W_z = v_y$ for all $S(z) = y$ minimizes this convex function. The corresponding Lagrangian is

$$\sum_{y=1}^{K} \sum_{z\in S_y} -f_y^\top W_z + m \sum_{y=1}^{K} \log \left(\sum_{i=1}^{C} \exp(f_y^\top W_i) \right) + \sum_{y=1}^{K} \nu_y(\|f_y\|_2^2 - 1) + \sum_{i=1}^{C} \lambda_i(\|W_i\|_2^2 - 1).$$

The stationarity condition with respect to W_z is the same as (A6), and we have already demonstrated that the feasibility constraints and complementary slackness are satisfied on W. The stationarity condition with respect to f_y is

$$-\sum_{z\in S_y} W_z + m \cdot \frac{\sum_{i=1}^{C} W_i \exp(f_y^\top W_i)}{\sum_{i=1}^{C} \exp(f_y^\top W_i)} + 2\lambda_y f_y = 0.$$

Substituting in $W_i = v_{S(i)}$ and $f_y = v_y$, we get $-\sum_{z\in S_y} v_y + m \cdot \frac{\sum_{i=1}^{C} v_{S(i)} \exp(v_y^\top v_{S(i)})}{\sum_{i=1}^{C} \exp(v_y^\top v_{S(i)})} + 2\lambda_y v_y = 0$. From the regular simplex definition, this is $-mv_y + m\frac{mv_y \exp(1) - mv_y \exp(\delta)}{m\exp(1) + (C-m)\exp(\delta)} + 2\lambda_y v_y = 0$. We thus have that $\lambda_y = \frac{m}{2}\left(1 - \frac{m(\exp(1) - \exp(\delta))}{m\exp(1) + (C-m)\exp(\delta)}\right)$, and the feasibility constraints are satisfied. Therefore, $f_y = W_z = v_y$ for $y = S(z)$ minimizes the generalization error $\mathcal{L}(x,z,f)$ when $p(h(x)|x) = 1$ and $p(z|x) = \frac{1}{m}$.

$p(z|x) = \frac{1}{m}$ and $p(y = h(x)|x) = 1$, so $p(z) = \int_{x:h(x)=S(z)} p(z,x)dx = \frac{1}{m} \int_{x:h(x)=S(z)} p(x) = \frac{1}{mK} = \frac{1}{C}$. $p(z)$ being class balanced means that $p(x|z) = \frac{p(z|x)p(x)}{p(z)} = Kp(x) = \frac{p(y|x)p(x)}{p(y)} = p(x|y)$. Therefore, this condition suggests that there is no distinction among the strata within a class. □

Proof of Observation 2. This observation follows directly from Observation 1 by repeating the proof approach with $z = y, m = 1$.

Lastly, suppose it is not true that $p(y = h(x)|x) = 1$. Then, the generalization error on the original task is $\mathcal{L}(x,y,f) = -\int_{\mathcal{X}} \sum_{y=1}^{K} p(x)p(y|x) \log \hat{p}(y|f(x))$, which is minimized when $\hat{p}(y|f(x)) = p(y|x)$. Intuitively, a model constructed with label information, $\hat{p}(y|h(x))$, will not improve over one that uses x itself to approximate $p(y|x)$. $\square$

Appendix D.2. Proofs for Theoretical Implications

We provide proofs for Section 4.3.

Proof of Lemma 1. The generalization error is

$$\begin{aligned}\mathcal{L}(x,z,\hat{f}_1) &= -\mathbb{E}_z\left[\mathbb{E}_{x\sim\mathcal{P}_z}\left[\log\frac{\exp(\hat{f}_1(x)^\top W_z)}{\sum_{i=1}^C \exp(\hat{f}_1(x)^\top W_i)}\right]\right]\\ &= \mathbb{E}_z\left[\mathbb{E}_{x\sim\mathcal{P}_z}\left[-\hat{f}_1(x)^\top W_z + \log\sum_{i=1}^C \exp(\hat{f}_1(x)^\top W_i)\right]\right].\end{aligned}$$

Using the definition of the mean classifier,

$$\begin{aligned}\mathcal{L}(x,z,\hat{f}_1) &= \mathbb{E}_z\left[-1+\mathbb{E}_{x\sim\mathcal{P}_z}\left[\log\sum_{i=1}^C \exp\left(\hat{f}_1(x)^\top \mathbb{E}_{x\sim\mathcal{P}_i}[\hat{f}_1(x)]\right)\right]\right]\\ &= -1+\mathbb{E}_z\left[\mathbb{E}_{x\sim\mathcal{P}_z}\left[\log\sum_{i=1}^C \exp\left(\hat{f}_1(x)^\top \mathbb{E}_i[\hat{f}_1(x)]\right)\right]\right].\end{aligned}$$

Since $\hat{f}_1(x)$ is bounded, there exists a constant $\lambda > 0$ such that

$$\mathbb{E}_{x\sim\mathcal{P}_z}\left[\log\sum_{i=1}^C \exp\left(\hat{f}_1(x)^\top \mathbb{E}_i[\hat{f}_1(x)]\right)\right] \leq \log\left(\sum_{i=1}^C \exp\left(\lambda\mathbb{E}_z[\hat{f}_1(x)]^\top \mathbb{E}_i[\hat{f}_1(x)]\right)\right).$$

We can also rewrite the dot product between mean embeddings per strata in terms of the distance between them:

$$\begin{aligned}\mathcal{L}(x,z,\hat{f}_1) &\leq -1+\mathbb{E}_z\left[\log\left(\sum_{i=1}^C \exp\left(\lambda\mathbb{E}_z[\hat{f}_1(x)]^\top \mathbb{E}_i[\hat{f}_1(x)]\right)\right)\right]\\ &= -1+\mathbb{E}_z\left[\log\left(\sum_{i=1}^C \exp\left(-\frac{\lambda}{2}\|\mathbb{E}_z[\hat{f}_1(x)] - \mathbb{E}_i[\hat{f}_1(x)]\|^2 + \lambda\right)\right)\right].\end{aligned}$$

This directly gives us our desired bound. $\square$

Proof of Lemma 2. The generalization error is

$$\begin{aligned}\mathcal{L}(x,y,\hat{f}_1) &= -\mathbb{E}_z\left[\mathbb{E}_{x\sim\mathcal{P}_z}\left[\log\frac{\exp(\hat{f}_1(x)^\top W_{S(z)})}{\sum_{i=1}^K \exp(\hat{f}_1(x)^\top W_i)}\right]\right]\\ &= \mathbb{E}_z\left[\mathbb{E}_{x\sim\mathcal{P}_z}\left[-\hat{f}_1(x)^\top W_{S(z)} + \log\sum_{i=1}^K \exp(\hat{f}_1(x)^\top W_i)\right]\right].\end{aligned}$$

We substitute in the definition of the mean classifier to get

$$\begin{aligned}\mathcal{L}(x,y,\hat{f}_1) = \mathbb{E}_z\Bigg[&-\sum_{z'\in S_{S(z)}} p(z'|S(z))\mathbb{E}_z[\hat{f}_1(x)]^\top \mathbb{E}_{z'}[\hat{f}_1(x)]\\ &+\mathbb{E}_{x\sim\mathcal{P}_z}\left[\log\sum_{i=1}^K \exp\left(\sum_{z'\in S_i} p(z'|S_i)\hat{f}_1(x)^\top \mathbb{E}_{z'}[\hat{f}_1(x)]\right)\right]\Bigg].\end{aligned}$$

We can rewrite the dot product between mean embeddings per strata in terms of the distance between them:

$$\begin{aligned}\mathcal{L}(x,y,\hat{f}_1) =&\mathbb{E}_z\Big[\sum_{z'\in S_{S(z)}} p(z'|S(z))\cdot\Big(\frac{1}{2}||\mathbb{E}_z[\hat{f}_1(x)]-\mathbb{E}_{z'}[\hat{f}_1(x)]||^2-1\Big)\\&+\mathbb{E}_{x\sim\mathcal{P}_z}\Big[\log\sum_{i=1}^{K}\exp\Big(\sum_{z'\in S_i}p(z'|S_i)\hat{f}_1(x)^\top\mathbb{E}_{z'}[\hat{f}_1(x)]\Big)\Big]\Big].\end{aligned}$$

We can write $||\mathbb{E}_z[\hat{f}_1(x)]-\mathbb{E}_{z'}[\hat{f}_1(x)]||$ in the above expression as $\delta(\hat{f}_1,z,z')$, which we have analyzed:

$$\begin{aligned}\mathcal{L}(x,y,\hat{f}_1) =&\mathbb{E}_z\Big[\sum_{z'\in S_{S(z)}} p(z'|S(z))\cdot\Big(\frac{1}{2}\delta(\hat{f}_1,z,z')^2-1\Big)\\&+\mathbb{E}_{x\sim\mathcal{P}_z}\Big[\log\sum_{i=1}^{K}\exp\Big(\sum_{z'\in S_i}p(z'|S_i)\hat{f}_1(x)^\top\mathbb{E}_{z'}[\hat{f}_1(x)]\Big)\Big]\Big].\end{aligned}$$

From our previous proof, there exists $\lambda>0$ such that this is at most

$$\begin{aligned}\mathcal{L}(x,y,\hat{f}_1) \leq&\mathbb{E}_z\Big[\sum_{z'\in S_{S(z)}} p(z'|S(z))\cdot\Big(\frac{1}{2}\delta(\hat{f}_1,z,z')^2-1\Big)\\&+\log\Big(\sum_{i=1}^{K}\exp\big(\sum_{z'\in S_i}p(z'|S_i)\lambda\mathbb{E}_z[\hat{f}_1(x)]^\top\mathbb{E}_{z'}[\hat{f}_1(x)]\big)\Big)\Big]\\=&\mathbb{E}_z\Big[\sum_{z'\in S_{S(z)}} p(z'|S(z))\cdot\Big(\frac{1}{2}\delta(\hat{f}_1,z,z')^2-1\Big)\\&+\log\Big(\sum_{i=1}^{K}\exp\Big(\sum_{z'\in S_i}p(z'|S_i)\Big(-\frac{\lambda}{2}||\mathbb{E}_z[\hat{f}_1(x)]-\mathbb{E}_{z'}[\hat{f}_1(x)]||^2+\lambda\Big)\Big)\Big)\Big].\end{aligned}$$

We can write each weighted summation over $p(z'|S(z))$ and $p(z'|S_i)$ as an expectation and use the definition of $\delta(\hat{f}_1,z,z')$ to obtain our desired bound. □

Appendix E. Additional Experimental Details

Appendix E.1. Datasets

We first describe all the datasets in more detail:

- **CIFAR10**, **CIFAR100**, and **MNIST** are all the standard computer vision datasets.
- **CIFAR10-Coarse** consists of two superclasses: animals (dog, cat, deer, horse, frog, bird) and vehicles (car, truck, plane, boat).
- **CIFAR100-Coarse** consists of twenty superclasses. We artificially imbalance subclasses to create **CIFAR100-Coarse-U**. For each superclass, we select one subclass to keep all 500 points, select one subclass to subsample to 250 points, select one subclass to subsample to 100 points, and select the remaining two to subsample to 50 points. We use the original CIFAR100 class index to select which subclasses to subsample: the subclass with the lowest original class index keeps all 500 points, the next subclass keeps 250 points, etc.
- **MNIST-Coarse** consists of two superclasses: <5 and ≥ 5.
- **Waterbirds** [14] is a robustness dataset designed to evaluate the effects of spurious correlations on model performance. The waterbirds dataset is constructed by cropping out birds from photos in the Caltech-UCSD Birds dataset [43], and pasting them on backgrounds from the Places dataset [44]. It consists of two categories: water birds and land birds. The water birds are heavily correlated with water backgrounds and the

land birds with land backgrounds, but 5% of the water birds are on land backgrounds, and 5% of the land birds are on water backgrounds. These form the (imbalanced) hidden strata.

- **ISIC** is a public skin cancer dataset for classifying skin lesions [15] as malignant or benign. 48% of the benign images contain a colored patch, which form the hidden strata.
- **CelebA** is an image dataset commonly used as a robustness benchmark [14,16]. The task is blonde/not blonde classification. Only 6% of blonde faces are male, which creates a rare stratum in the blonde class.

Appendix E.2. Hyperparameters

For all model quality experiments for L_{spread}, we first fixed $\tau = 0.5$ and swept $\alpha \in [0.16, 0.25, 0.33, 0.5, 0.67]$. We then took the two best-performing values and swept $\tau \in [0.1, 0.3, 0.5, 0.7, 0.9]$. For L_{SC} and L_{SS}, we swept $\tau \in [0.1, 0.3, 0.5, 0.7, 0.9]$. Final hyperparameter values for (τ, α) for L_{spread} were $(0.9, 0.67)$ for CIFAR10, $(0.5, 0.16)$ for CIFAR10-coarse, $(0.5, 0.33)$ for CIFAR100, $(0.5, 0.25)$ for CIFAR100-Coarse, $(0.5, 0.25)$ for CIFAR100-Coarse-U, $(0.5, 0.5)$ for MNIST, $(0.5, 0.5)$ for MNIST-coarse, $(0.5, 0.5)$ for ISIC, and $(0.5, 0.5)$ for waterbirds.

For coarse-to-fine transfer learning, we fixed $\tau = 0.5$ for all losses and swept $\alpha \in [0.16, 0.25, 0.33, 0.5, 0.67]$. Final hyperparameter values for α were 0.25 for CIFAR10-Coarse, 0.25 for CIFAR100-Coarse, 0.25 for CIFAR100-Coarse-U, and 0.5 for MNIST-Coarse.

Appendix E.3. Applications

We describe additional experimental details for the applications.

Appendix E.3.1. Robustness Against Worst-Group Performance

We follow the evaluation of [5]. First, we train a model on the standard class labels. We evaluate different loss functions for this step, including L_{spread}, L_{SC}, and the cross entropy loss L_{CE}. Then we project embeddings of the training set using a UMAP projection [45], and cluster points to discover unlabeled subgroups. Finally, we use the unlabeled subgroups in a Group-DRO algorithm to optimize worst-group robustness [14].

Appendix E.3.2. Robustness Against Noise

We use the same training setup as we use to evaluate model quality, and introduce symmetric noise into the labels for the contrastive loss head. We train the cross entropy head with a fraction of the full training set. In Section 5.3, we report results from training with 20% labels to cross entropy. We report additional levels in Appendix F.

We detect noisy labels with a simple geometric heuristic: for each point, we compute the cosine similarity between the embedding of the point and the center of all the other points in the batch that have the same class. We compare this similarity value to the average cosine similarity with points in the batch from every other class, and rank the points by the difference between these two values. Points with incorrect labels have a small difference between these two values (they appear to be small strata, so they are far away from points of the same class). Given the noise level ϵ as an input, we rank the points by this heuristic and mark the ϵ fraction of the batch with the smallest scores as noisy. We then correct their labels by adopting the label of the closest cluster center.

Appendix E.3.3. Minimal Coreset Construction

We use the publicly-available evaluation framework for coresets from [18] (https://github.com/mtoneva/example_forgetting, accessed on 1 October 2021). We use the official repository from [19] (https://github.com/mansheej/data_diet, accessed on 1 October 2021) to recreate their coreset algorithms.

Our coreset algorithm proceeds in two parts. First, we give each point a difficulty rating based on how likely we are to classify it correctly under partial training. Then we subsample the easiest points to construct minimal coresets.

First, we mirror the set up from our thought experiment and train with L_{spread} on random samples of $t\%$ of the CIFAR10 training set, taking three random samples for each of $t \in [10, 20, 50]$ (and we train the cross entropy head with 1% labeled data). For each run, we record which points are classified correctly by the cross entropy head at the end of training, and bucket points the training set by how often the point was correctly classified. To construct a coreset of size $t\%$, we iteratively remove points from the largest bucket in each class. Our strategy removes easy examples first from the largest coresets, but maintains a set of easy examples in the smallest coresets.

Appendix F. Additional Experimental Results

In this section, we report three sets of additional experimental results: the performance of using $L_{attract}$ on its own to train models, sample complexity of L_{spread} compared to L_{SC}, and additional noisy label results (including a bonus de-noising algorithm).

Appendix F.1. Performance of $L_{attract}$

In an early iteration of this project, we experienced success with using $L_{attract}$ on its own to train models, before realizing the benefits of adding in an additional term to prevent class collapse. As an ablation, we report on the performance of using $L_{attract}$ on its own in Table A2. $L_{attract}$ can outperform L_{SC}, but L_{spread} outperforms both. We do not report the results here, but $L_{attract}$ also performs significantly worse than L_{SC} on downstream applications, since it more direclty encourages class collapse.

Table A2. Performance of L_{spread} compared to L_{SC} and using $L_{attract}$ on its own. Best in bold.

	End Model Perf.			
Dataset	L_{SS}	L_{SC}	$L_{attract}$	L_{spread}
CIFAR10	89.7	90.9	91.3	**91.5**
CIFAR100	68.0	67.5	68.9	**69.1**

Appendix F.2. Sample Complexity

Figure A1 shows the performance of training ViT models with various amounts of labeled data for L_{spread}, L_{SC}, and L_{SS}. In these experiments, we train the cross entropy head with 1% labeled data to isolate the effect of training data on the contrastive losses themselves.

L_{spread} outperforms L_{SC} and L_{SS} throughout. At 10% labeled data, L_{spread} outperforms L_{SS} by 13.9 points, and outperforms L_{SC} by 0.5 points. By 100% labeled data (for the contrastive head), L_{spread} outperforms L_{SS} by 25.4 points, and outperforms L_{SC} by 10.3 points.

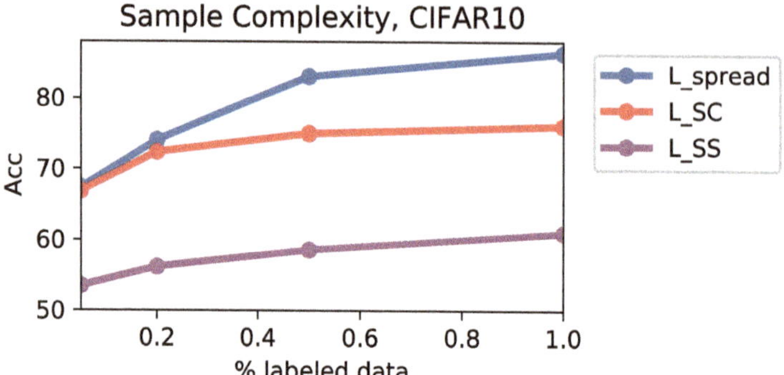

Figure A1. Performance of training ViT with L_{spread} compared to training with L_{SC} and L_{SS} on CIFAR10 at various amounts of labeled data. L_{spread} outperforms the baselines at each point. The cross entropy head here is trained with 1% labeled data to isolate the effect of training data on the contrastive losses.

Appendix F.3. Noisy Labels

In Section 5.3, we reported results from training the contrastive loss head with noisy labels and the cross entropy loss with clean labels from 20% of the training data.

In this section, we first discuss a de-noising algorithm inspired by [23] that we initially developed to correct for noisy labels, but that we did not observe strong empirical results from. We hope that reporting this result inspires future work into improving contrastive learning.

We then report additional results with larger amounts of training data for the cross entropy head.

Appendix F.3.1. Debiasing Noisy Contrastive Loss

First, we consider the triplet loss and show how to debias it in expectation under noise. Then we present an extension to supervised contrastive loss.

Noise-Aware Triplet Loss

Consider the triplet loss:

$$L_{triplet} = \mathbb{E}_{\substack{x \sim \mathcal{P}, x^+ \sim p^+(\cdot|x), \\ x^- \sim p^-(\cdot|x)}} \left[-\log \frac{\exp(\sigma(x, x^+))}{\exp(\sigma(x, x^+)) + \exp(\sigma(x, x^-))} \right]. \tag{A7}$$

Now suppose that we do not have access to true labels but instead have noisy labels denoted by the weak classifier $\widetilde{y} := \widetilde{h}(x)$. We adopt a simple model of symmetric noise where $\widetilde{p} = \Pr(\text{noisy label is correct})$.

We use $\widetilde{y}$ to construct $\widetilde{\mathcal{P}}^+$ and $\widetilde{\mathcal{P}}^-$ as $p(x^+|\widetilde{h}(x) = \widetilde{h}(x^+))$ and $p(x^-|\widetilde{h}(x) \neq \widetilde{h}(x^-))$. For simplicity, we start by looking at how the triplet loss in (A7) is impacted when *noise is not addressed* in the binary setting. Define $L_{noisy}^{triplet}$ as $L_{triplet}$ used with $\widetilde{\mathcal{P}}^+$ and $\widetilde{\mathcal{P}}^-$.

Lemma A3. *When class-conditional noise is uncorrected,* $L_{triplet}^{noisy}$ *is equivalent to*

$$(\widetilde{p}^3 + (1-\widetilde{p})^3)L_{triplet} + \widetilde{p}(1-\widetilde{p})\mathbb{E}_{\substack{x\sim\mathcal{P} \\ x_1^+,x_2^+\sim p^+(\cdot|x)}}\left[-\log\frac{\exp(\sigma(x,x_1^+))}{\exp(\sigma(x,x_1^+))+\exp(\sigma(x,x_2^+))}\right]$$
$$+ \widetilde{p}(1-\widetilde{p})\mathbb{E}_{\substack{x\sim\mathcal{P} \\ x_1^-,x_2^-\sim p^-(\cdot|x)}}\left[-\log\frac{\exp(\sigma(x,x_1^-))}{\exp(\sigma(x,x_1^-))+\exp(\sigma(x,x_2^-))}\right]$$
$$+ \widetilde{p}(1-\widetilde{p})\mathbb{E}_{\substack{x\sim\mathcal{P} \\ x^+\sim p^+(\cdot|x) \\ x^-\sim p^-(\cdot|x)}}\left[-\log\frac{\exp(\sigma(x,x^-))}{\exp(\sigma(x,x^+))+\exp(\sigma(x,x^-))}\right].$$

Proof. We split $L_{triplet}^{noisy}$ depending on if the noisy positive and negative pairs are truly positive and negative.

$$L_{triplet}^{noisy} = \mathbb{E}_{\substack{x\sim\mathcal{P} \\ \widetilde{x}^+\sim\widetilde{p}^+(\cdot|x) \\ \widetilde{x}^-\sim\widetilde{p}^-(\cdot|x)}}\left[-\log\frac{\exp(\sigma(x,\widetilde{x}^+))}{\exp(\sigma(x,\widetilde{x}^+))+\exp(\sigma(x,\widetilde{x}^-))}\right]$$
$$= p(h(x)=h(\widetilde{x}^+),h(x)\neq h(\widetilde{x}^-))\mathbb{E}_{\substack{x\sim\mathcal{P} \\ x^+\sim p^+(\cdot|x) \\ x^-\sim p^-(\cdot|x)}}\left[-\log\frac{\exp(\sigma(x,x^+))}{\exp(\sigma(x,x^+))+\exp(\sigma(x,x^-))}\right]$$
$$+ p(h(x)=h(\widetilde{x}^+),h(x)=h(\widetilde{x}^-))\mathbb{E}_{\substack{x\sim\mathcal{P} \\ x_1^+,x_2^+\sim p^+(\cdot|x)}}\left[-\log\frac{\exp(\sigma(x,x_1^+))}{\exp(\sigma(x,x_1^+))+\exp(\sigma(x,x_2^+))}\right]$$
$$+ p(h(x)\neq h(\widetilde{x}^+),h(x)\neq h(\widetilde{x}^-))\mathbb{E}_{\substack{x\sim\mathcal{P} \\ x_1^-,x_2^-\sim p^-(\cdot|x)}}\left[-\log\frac{\exp(\sigma(x,x_1^-))}{\exp(\sigma(x,x_1^-))+\exp(\sigma(x,x_2^-))}\right]$$
$$+ p(h(x)\neq h(\widetilde{x}^+),h(x)=h(\widetilde{x}^-))\mathbb{E}_{\substack{x\sim\mathcal{P} \\ x^+\sim p^+(\cdot|x) \\ x^-\sim p^-(\cdot|x)}}\left[-\log\frac{\exp(\sigma(x,x^-))}{\exp(\sigma(x,x^+))+\exp(\sigma(x,x^-))}\right].$$

Define $\widetilde{p} = p(\text{noisy label is correct})$. Note that

$$p(h(x)=h(\widetilde{x}^+),h(x)\neq h(\widetilde{x}^-)) = \widetilde{p}^3 + (1-\widetilde{p})^3,$$

(i.e., all three points are correct or all reversed, such that their relative pairings are correct). In addition, the other three probabilities above are all equal to $\widetilde{p}(1-\widetilde{p})$. □

We now show that there exists a weighted loss function that in expectation equals $L_{triplet}$.

Lemma A4. *Define*

$$\widetilde{L}_{triplet} = \mathbb{E}_{\substack{x\sim\mathcal{P}, \\ \widetilde{x}_1^+,\widetilde{x}_2^+\sim\widetilde{\mathcal{P}}^+(\cdot|x) \\ \widetilde{x}_1^-,\widetilde{x}_2^-\sim\widetilde{\mathcal{P}}^-(\cdot|x)}}\Big[-w^+\sigma(x,\widetilde{x}_1^+) + w^-\sigma(x,\widetilde{x}_1^-)$$
$$+ w_1\log\left(\exp\left(\sigma(x,\widetilde{x}_1^+)\right)+\exp\left(\sigma(x,\widetilde{x}_1^-)\right)\right)$$
$$- w_2\log\Big((\exp(\sigma(x,\widetilde{x}_1^+))+\exp(\sigma(x,\widetilde{x}_2^+)))\cdot(\exp(\sigma(x,\widetilde{x}_1^-))+\exp(\sigma(x,\widetilde{x}_2^-)))\Big)\Big],$$

where

$$w^+ = \frac{\widetilde{p}^2 + (1-\widetilde{p})^2}{(2\widetilde{p}-1)^2} \qquad w^- = \frac{2\widetilde{p}(1-\widetilde{p})}{(2\widetilde{p}-1)^2} \qquad w_1 = \frac{\widetilde{p}^2 + (1-\widetilde{p})^2}{(2\widetilde{p}-1)^2} \qquad w_2 = \frac{\widetilde{p}(1-\widetilde{p})}{(2\widetilde{p}-1)^2}.$$

Then, $\mathbb{E}\left[\widetilde{L}_{triplet}\right] = L_{triplet}$.

Proof. We evaluate $\mathbb{E}\left[-w_1\sigma(x,\widetilde{x}_1^+) + w_2\sigma(x,\widetilde{x}_1^-)\right]$ and the other terms separately. Using the same probabilities as computed in Lemma A3,

$$\begin{aligned}
&\mathbb{E}\left[-w_1\sigma(x,\widetilde{x}_1^+) + w_2\sigma(x,\widetilde{x}_1^-)\right] = -(\widetilde{p}^2 + (1-\widetilde{p})^2)w_1\mathbb{E}\left[\sigma(x,x_1^+)\right] \\
&- 2\widetilde{p}(1-\widetilde{p})w_1\mathbb{E}\left[\sigma(x,x_1^-)\right] + (\widetilde{p}^2 + (1-\widetilde{p})^2)w_2\mathbb{E}\left[\sigma(x,x_1^-)\right] + 2\widetilde{p}(1-\widetilde{p})w_2\mathbb{E}\left[\sigma(x,x_1^+)\right] \\
&= -\mathbb{E}\left[\sigma(x,x_1^+)\right].
\end{aligned}$$

We evaluate the remaining terms:

$$\begin{aligned}
&\mathbb{E}\left[w_3\log\left(\exp\left(\sigma(x,\widetilde{x}_1^+)\right) + \exp\left(\sigma(x,\widetilde{x}_1^-)\right)\right)\right] = \\
&(\widetilde{p}^2 + (1-\widetilde{p})^2)w_3\mathbb{E}\left[\log\left(\exp\left(\sigma(x,x_1^+)\right) + \exp\left(\sigma(x,x_1^-)\right)\right)\right] \\
&+ \widetilde{p}(1-\widetilde{p})w_3\mathbb{E}\left[\log\left((\exp(\sigma(x,\widetilde{x}_1^+)) + \exp(\sigma(x,\widetilde{x}_2^+)))\cdot(\exp(\sigma(x,\widetilde{x}_1^-)) + \exp(\sigma(x,\widetilde{x}_2^-)))\right)\right].
\end{aligned}$$

and

$$\begin{aligned}
&\mathbb{E}\left[w_4\log\left(\exp\left(\sigma(x,\widetilde{x}_1^+)\right) + \exp\left(\sigma(x,\widetilde{x}_2^+)\right)\right)\right] \\
&+ \mathbb{E}\left[w_4\log\left(\exp\left(\sigma(x,\widetilde{x}_1^-)\right) + \exp\left(\sigma(x,\widetilde{x}_2^-)\right)\right)\right] = \\
&(\widetilde{p}^2 + (1-\widetilde{p})^2)w_4\mathbb{E}\left[\log\left(\exp\left(\sigma(x,x_1^+)\right) + \exp\left(\sigma(x,x_2^+)\right)\right)\right] \\
&+ 4\widetilde{p}(1-\widetilde{p})w_4\mathbb{E}\left[\log\left(\exp\left(\sigma(x,x_1^+)\right) + \exp\left(\sigma(x,x_1^-)\right)\right)\right] \\
&+ ((1-\widetilde{p})^2 + \widetilde{p}^2)w_4\mathbb{E}\left[\log\left(\exp\left(\sigma(x,x_1^-)\right) + \exp\left(\sigma(x,x_2^-)\right)\right)\right].
\end{aligned}$$

Examining the coefficients, we see that

$$\begin{aligned}
(\widetilde{p}^2 + (1-\widetilde{p})^2)w_3 - 4\widetilde{p}(1-\widetilde{p})w_4 &= \frac{(\widetilde{p}^2 + (1-\widetilde{p})^2)^2}{(2\widetilde{p}-1)^2} - \frac{4\widetilde{p}^2(1-\widetilde{p})^2}{(2\widetilde{p}-1)^2} = 1 \\
\widetilde{p}(1-\widetilde{p})w_3 - (\widetilde{p}^2 + (1-\widetilde{p})^2)w_4 &= \frac{\widetilde{p}(1-\widetilde{p})(\widetilde{p}^2 + (1-\widetilde{p})^2)}{(2\widetilde{p}-1)^2} - \frac{(\widetilde{p}^2 + (1-\widetilde{p})^2)\widetilde{p}(1-\widetilde{p})}{(2\widetilde{p}-1)^2} = 0,
\end{aligned}$$

which shows that only the term $\mathbb{E}\left[\log\left(\exp\left(\sigma(x,x_1^+)\right) + \exp\left(\sigma(x,x_1^-)\right)\right)\right]$ persists. This completes our proof. □

We now show the general case for debiasing $L_{attract}$, which uses more negative samples.

Proposition A1. *Define* $m = n+1$ *(as the "batch size" in the denominator), and*

$$\widetilde{L}_{attract} = \mathbb{E}_{\substack{x\sim\mathcal{P} \\ \{\widetilde{x}_i^+\}_{i=1}^m \\ \{\widetilde{x}_j^-\}_{j=1}^m}}\Bigg[-w^+\sigma(x,\widetilde{x}_1^+) + w^-\sigma(x,\widetilde{x}_1^-) \tag{A8}$$

$$+ \sum_{k=0}^{m} w_k\log\left(\sum_{i=1}^{k}\exp\left(\sigma(x,\widetilde{x}_i^+)\right) + \sum_{j=1}^{m-k}\exp\left(\sigma(x,\widetilde{x}_j^-)\right)\right)\Bigg]. \tag{A9}$$

w^+ and w^- are defined in the same was as before. $\vec{w} = \{w_0, \dots w_m\} \in \mathbb{R}^{m+1}$ is the solution to the system $\mathbf{P}w = \mathbf{e}_2$ where $\mathbf{e}_2$ is the standard basis vector in $\mathbb{R}^{m+1}$ where the 2nd index is 1 and all others are 0. The i, jth element of $\mathbf{P}$ is $\mathbf{P}_{ij} = \widetilde{p}\mathbf{Q}_{i,j} + (1 - \widetilde{p})\mathbf{Q}_{m-i,j}$ where

$$\mathbf{Q}_{i,j} = \begin{cases} \sum_{k=0}^{\min\{j,m-i\}} \binom{j}{k}\binom{m-j}{i-j+k}(1-\widetilde{p})^{i-j+2k}\widetilde{p}^{m+j-i-2k} & j \leq i \\ \sum_{k=0}^{\min\{i,m-j\}} \binom{m-j}{k}\binom{j}{j-i+k}(1-\widetilde{p})^{j-i+2k}\widetilde{p}^{m-j+i-2k} & j > i \end{cases}$$

Then, $\mathbb{E}\left[\widetilde{L}_{attract}\right] = L_{attract}$.

We do not present the proof for Proposition A1, but the steps are very similar to the proof for the triplet loss case. We also note that a different form of $\mathbb{E}\left[\widetilde{L}_{attract}\right]$ must be computed for the multi-class case, which we do not present here (but can be derived through computation).

Observation 4. *Note that the values of $\mathbf{Q}_{\mathbf{i},\mathbf{j}}$ have high variance in the noise rate as m increases. Additionally, note that the number of terms in the summation of $\mathbf{Q}_{\mathbf{i},\mathbf{j}}$ increase combinatorially with m. We found this de-noising algorithm very unstable as a result.*

Appendix F.3.2. Additional Noisy Label Results

Now we report the performance of denoising algorithms with additional amounts of labeled data for the cross entropy loss head. We also report the performance of using $\widetilde{L}_{attract}$ to debias noisy labels.

Figure A2 shows the results. Our geometric correction together with L_{spread} works the most consistently. Using the geometric correction with L_{SC} can be unreliable, since L_{SC} can learn memorize noisy labels early on in training. The expectation-based debiasing algorithm $\widetilde{L}_{attract}$ occasionally shows promise but is unreliable, and is very sensitive to having the correct noise rate as an input.

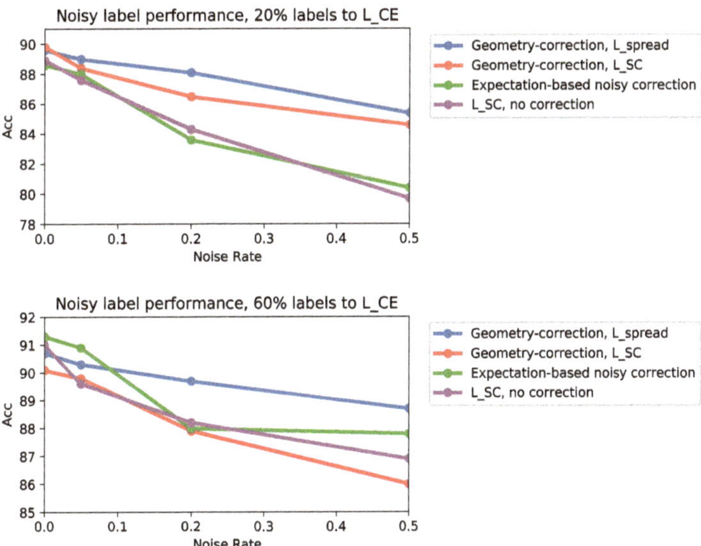

Figure A2. Performance of models under various amounts of label noise for the contrastive loss head, and various amounts of clean training data for the cross entropy loss.

References

1. Khosla, P.; Teterwak, P.; Wang, C.; Sarna, A.; Tian, Y.; Isola, P.; Mschinot, A.; Liu, C.; Krishnan, D. Supervised Contrastive Learning. *Adv. Neural Inf. Process. Syst.* **2020**, *33*, 18661–18673.
2. Graf, F.; Hofer, C.; Niethammer, M.; Kwitt, R. Dissecting Supervised Constrastive Learning. *Proc. Int. Conf. Mach. Learn. PMLR.* **2021**, *139*, 3821–3830.
3. Zhang, C.; Bengio, S.; Hardt, M.; Recht, B.; Vinyals, O. Understanding deep learning requires rethinking generalization. *Commun. ACM* **2016**, *64*, 107–115. [CrossRef]
4. Hoffmann, A.; Kwok, R.; Compton, P. Using subclasses to improve classification learning. In *European Conference on Machine Learning*; Springer: Berlin/Heidelberg, Germany, 2001; pp. 203–213.
5. Sohoni, N.; Dunnmon, J.; Angus, G.; Gu, A.; Ré, C. No Subclass Left Behind: Fine-Grained Robustness in Coarse-Grained Classification Problems. *Adv. Neural Inf. Process. Syst.* **2020**, *33*, 19339–19352.
6. Oakden-Rayner, L.; Dunnmon, J.; Carneiro, G.; Ré, C. Hidden stratification causes clinically meaningful failures in machine learning for medical imaging. In Proceedings of the Proceedings of the ACM conference on health, inference, and learning, Toronto, ON, Canada, 2–4 April 2020; pp. 151–159.
7. Linsker, R. Self-organization in a perceptual network. *Computer* **1988**, *21*, 105–117. [CrossRef]
8. Wang, T.; Isola, P. Understanding contrastive representation learning through alignment and uniformity on the hypersphere. *Proc. Int. Conf. Mach. Learn. PMLR* **2020**, *119*, 9929–9939.
9. Robinson, J.; Chuang, C.Y.; Sra, S.; Jegelka, S. Contrastive learning with hard negative samples. *arXiv* **2020**, arXiv:2010.04592.
10. Ben-David, S.; Blitzer, J.; Crammer, K.; Kulesza, A.; Pereira, F.; Vaughan, J.W. A theory of learning from different domains. *Mach. Learn.* **2010**, *79*, 151–175. [CrossRef]
11. Ben-David, S.; Blitzer, J.; Crammer, K.; Pereira, F. Analysis of representations for domain adaptation. *Adv. Neural Inf. Process. Syst.* **2007**, *19*, 137.
12. Ben-David, S.; Urner, R. On the Hardness of Domain Adaptation and the Utility of Unlabeled Target Samples. In Proceedings of the 23rd International Conference, Lyon, France, 29–31 October 2012; Bshouty, N.H., Stoltz, G., Vayatis, N., Zeugmann, T., Eds.; Springer: Berlin/Heidelberg, Germany, 2012; pp. 139–153.
13. Mansour, Y.; Mohri, M.; Rostamizadeh, A. Domain adaptation: Learning bounds and algorithms. *arXiv* **2009**, arXiv:0902.3430.
14. Sagawa, S.; Koh, P.W.; Hashimoto, T.B.; Liang, P. Distributionally Robust Neural Networks for Group Shifts: On the Importance of Regularization for Worst-Case Generalization. *arXiv* **2019**, arXiv:1911.08731.
15. Codella, N.; Rotemberg, V.; Tschandl, P.; Celebi, M.E.; Dusza, S.; Gutman, D.; Helba, B.; Kalloo, A.; Liopyris, K.; Marchetti, M.; et al. Skin lesion analysis toward melanoma detection 2018: A challenge hosted by the international skin imaging collaboration (isic). *arXiv* **2019**, arXiv:1902.03368.
16. Liu, Z.; Luo, P.; Wang, X.; Tang, X. Deep Learning Face Attributes in the Wild. In Proceedings of the Proceedings of International Conference on Computer Vision (ICCV), Santiago, Chile, 7 December 2015.
17. Dosovitskiy, A.; Beyer, L.; Kolesnikov, A.; Weissenborn, D.; Zhai, X.; Unterthiner, T.; Dehghani, M.; Minderer, M.; Heigold, G.; Gelly, S.; et al. An Image is Worth 16x16 Words: Transformers for Image Recognition at Scale. *arXiv* **2020**, arXiv:2010.11929.
18. Toneva, M.; Sordoni, A.; des Combes, R.T.; Trischler, A.; Bengio, Y.; Gordon, G.J. An Empirical Study of Example Forgetting during Deep Neural Network Learning. *arXiv* **2018**, arXiv:1812.05159.
19. Paul, M.; Ganguli, S.; Dziugaite, G.K. Deep Learning on a Data Diet: Finding Important Examples Early in Training. *arXiv* **2021**, arXiv:2107.07075.
20. Chen, T.; Kornblith, S.; Norouzi, M.; Hinton, G. A simple framework for contrastive learning of visual representations. *Proc. Int. Conf. Mach. Learn. PMLR* **2020**, *119*, 1597–1607
21. Arora, S.; Khandeparkar, H.; Khodak, M.; Plevrakis, O.; Saunshi, N. A theoretical analysis of contrastive unsupervised representation learning. *arXiv* **2019**, arXiv:1902.09229.
22. Zimmermann, R.S.; Sharma, Y.; Schneider, S.; Bethge, M.; Brendel, W. Contrastive Learning Inverts the Data Generating Process. *arXiv* **2021**, arXiv:2012.08850.
23. Chuang, C.Y.; Robinson, J.; Torralba, A.; Jegelka, S. Debiased Contrastive Learning. *Adv. Neural Inf. Process. Syst.* **2020**, *33*, 8765–8775.
24. Oord, A.v.d.; Li, Y.; Vinyals, O. Representation learning with contrastive predictive coding. *arXiv* **2018**, arXiv:1807.03748.
25. Tian, Y.; Sun, C.; Poole, B.; Krishnan, D.; Schmid, C.; Isola, P. What makes for good views for contrastive learning? *arXiv* **2020**, arXiv:2005.10243.
26. Tsai, Y.H.H.; Wu, Y.; Salakhutdinov, R.; Morency, L.P. Self-supervised Learning from a Multi-view Perspective. *arXiv* **2020**, arXiv:2006.05576.
27. Tschannen, M.; Djolonga, J.; Rubenstein, P.K.; Gelly, S.; Lucic, M. On Mutual Information Maximization for Representation Learning. *arXiv* **2019**, arXiv:1907.13625.
28. He, K.; Fan, H.; Wu, Y.; Xie, S.; Girshick, R. Momentum Contrast for Unsupervised Visual Representation Learning. *arXiv* **2019**, arXiv:1911.05722.
29. Chen, X.; Fan, H.; Girshick, R.; He, K. Improved Baselines with Momentum Contrastive Learning. *arXiv* **2020**, arXiv:2003.04297.
30. Goyal, P.; Caron, M.; Lefaudeux, B.; Xu, M.; Wang, P.; Pai, V.; Singh, M.; Liptchinsky, V.; Misra, I.; Joulin, A.; et al. Self-supervised Pretraining of Visual Features in the Wild. *arXiv* **2021**, arXiv:2103.01988.

31. Caron, M.; Misra, I.; Mairal, J.; Goyal, P.; Bojanowski, P.; Joulin, A. Unsupervised Learning of Visual Features by Contrasting Cluster Assignments. *Adv. Neural Inf. Process. Syst.* **2020**, *33*, 9912–9924.
32. Islam, A.; Chen, C.F.; Panda, R.; Karlinsky, L.; Radke, R.; Feris, R. A Broad Study on the Transferability of Visual Representations with Contrastive Learning. *arXiv* **2021**, arXiv:2103.13517.
33. Bukchin, G.; Schwartz, E.; Saenko, K.; Shahar, O.; Feris, R.; Giryes, R.; Karlinsky, L. Fine-grained Angular Contrastive Learning with Coarse Labels. In Proceedings of the 2021 IEEE/CVF Conference on Computer Vision and Pattern Recognition (CVPR), Virtual, 19 June 2021; [CrossRef]
34. d'Eon, G.; d'Eon, J.; Wright, J.R.; Leyton-Brown, K. The Spotlight: A General Method for Discovering Systematic Errors in Deep Learning Models. *arXiv* **2021**, arXiv:2107.00758.
35. Duchi, J.; Hashimoto, T.; Namkoong, H. Distributionally robust losses for latent covariate mixtures. *arXiv* **2020**, arXiv:2007.13982.
36. Goel, K.; Gu, A.; Li, Y.; Re, C. Model Patching: Closing the Subgroup Performance Gap with Data Augmentation. *arXiv* **2020**, arXiv:2008.06775.
37. Liu, S.; Niles-Weed, J.; Razavian, N.; Fernandez-Granda, C. Early-Learning Regularization Prevents Memorization of Noisy Labels. *Adv. Neural Inf. Process. Syst.* **2020**, *33*, 20331–20342.
38. Li, J.; Xiong, C.; Hoi, S.C. Semi-supervised Learning with Contrastive Graph Regularization. In Proceedings of the IEEE/CVF International Conference on Computer Vision (ICCV), Virtual, 11 October 2021.
39. Ciortan, M.; Dupuis, R.; Peel, T. A Framework using Contrastive Learning for Classification with Noisy Labels. *arXiv* **2021**, arXiv:2104.09563.
40. Li, J.; Socher, R.; Hoi, S.C. DivideMix: Learning with Noisy Labels as Semi-supervised Learning. *arXiv* **2020**, arXiv:2002.07394.
41. Ju, J.; Jung, H.; Oh, Y.; Kim, J. Extending Contrastive Learning to Unsupervised Coreset Selection. *arXiv* **2021**, arXiv:2103.03574.
42. Sener, O.; Savarese, S. Active Learning for Convolutional Neural Networks: A Core-Set Approach. *arXiv* **2017**, arXiv:1708.00489.
43. Welinder, P.; Branson, S.; Mita, T.; Wah, C.; Schroff, F.; Belongie, S.; Perona, P. Caltech-UCSD Birds 200. In *Technical Report CNS-TR-2010-001*; California Institute of Technology: Pasadena, CA, USA, 2010.
44. Zhou, B.; Lapedriza, A.; Xiao, J.; Torralba, A.; Oliva, A. Learning Deep Features for Scene Recognition using Places Database. *Adv. Neural Inf. Process. Syst.* **2014**, *27*, 487–495.
45. McInnes, L.; Healy, J.; Saul, N.; Großberger, L. UMAP: Uniform Manifold Approximation and Projection. *J. Open Source Softw.* **2018**, *3*. [CrossRef]

computer sciences and mathematics forum

Proceeding Paper

DAP-SDD: Distribution-Aware Pseudo Labeling for Small Defect Detection †

Xiaoyan Zhuo [1,*], Wolfgang Rahfeldt [2], Xiaoqian Zhang [2], Ted Doros [2] and Seung Woo Son [1]

1 Department of Electrical and Computer Engineering, University of Massachusetts Lowell, Lowell, MA 01854, USA; seungwoo_son@uml.edu
2 Data Science, Micron Technology, Inc., Manassas, VA 20110, USA; wrahfeldt@micron.com (W.R.); xiaoqianzhan@micron.com (X.Z.); tdoros@micron.com (T.D.)
* Correspondence: xiaoyan_zhuo@student.uml.edu

† Presented at the AAAI Workshop on Artificial Intelligence with Biased or Scarce Data (AIBSD), Online, 28 February 2022.

Abstract: Detecting defects, especially when they are small in the early manufacturing stages, is critical to achieving a high yield in industrial applications. While numerous modern deep learning models can improve detection performance, they become less effective in detecting small defects in practical applications due to the scarcity of labeled data and significant class imbalance in multiple dimensions. In this work, we propose a distribution-aware pseudo labeling method (DAP-SDD) to detect small defects accurately while using limited labeled data effectively. Specifically, we apply bootstrapping on limited labeled data and then utilize the approximated label distribution to guide pseudo label propagation. Moreover, we propose to use the t-distribution confidence interval for threshold setting to generate more pseudo labels with high confidence. DAP-SDD also incorporates data augmentation to enhance the model's performance and robustness. We conduct extensive experiments on various datasets to validate the proposed method. Our evaluation results show that, overall, our proposed method requires less than 10% of labeled data to achieve comparable results of using a fully-labeled (100%) dataset and outperforms the state-of-the-art methods. For a dataset of wafer images, our proposed model can achieve above 0.93 of AP (average precision) with only four labeled images (i.e., 2% of labeled data).

Keywords: pseudo labeling; small defect detection; t-distribution; threshold setting

Citation: Zhuo, X.; Rahfeldt, W.; Zhang, X.; Doros, T., Son, S.W. DAP-SDD: Distribution-Aware Pseudo Labeling for Small Defect Detection. *CSFM* **2022**, *3*, 5. https://doi.org/10.3390/cmsf2022003005

Academic Editors: Kuan-Chuan Peng and Ziyan Wu

Published: 20 April 2022

Publisher's Note: MDPI stays neutral with regard to jurisdictional claims in published maps and institutional affiliations.

1. Introduction

In the semiconductor industry, detecting small defects at the early stages of manufacturing is crucial for improving yield and saving costs. For example, as wafers are processed in batches or lots, malfunctioning tools or suboptimal operations may result in whole batches of wafers suffering mass yield loss or even being discarded [1,2]. If we can detect anomalies early, tool issues or operation problems can be fixed quickly before more batches of wafers travel through malfunctioning tools or undergo unnecessary value-adding manufacturing processes. Small defects on wafer images usually indicate early-phase tool malfunctions or improper operations. However, due to the high variance of working conditions (e.g., position, orientation, illumination) and complex calibration procedures [2], traditional inspection tools lack the flexibility to detect various defects and suffer from poor detection performance, especially for small and dim defects.

In recent years, numerous deep learning models for object detection have been proposed, such as object detection models [3–6] and segmentation models [7–9] and have demonstrated impressive improvements in detecting objects. However, they suffer from a performance bottleneck on detecting small objects [10,11] due to several factors. First, small objects have a limited number of pixels to represent information. Additionally, small objects are scarce in the training dataset [10,12]. Furthermore, key features that can be used to distinguish small objects from a background or other categories are vulnerable or even lost while going through deep layers of networks, such as convolution or pooling layers [13]. Figure 1 presents examples of small defects we explore in this work, which have these previously mentioned challenges. In these industrial inspection datasets, the sizes of defects range from 3 × 3 to 31 × 31 pixels and smaller than 16 × 16 on average. Moreover, there are usually fewer than four small defects in each image.

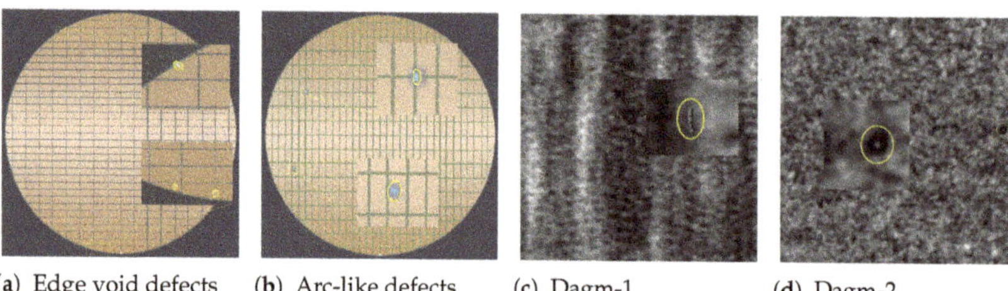

(**a**) Edge void defects (**b**) Arc-like defects (**c**) Dagm-1 (**d**) Dagm-2

Figure 1. Examples of small defects that we explored in this work. (**a**,**b**) are examples of in-house wafer image datasets; (**c**,**d**) are examples of industrial optical inspection [14]. Due to confidentiality reasons, the wafer images are artificially-created ones that approximate the real-world data for demonstration purposes; we use the real-world dataset for model training and evaluation in this work.

Several studies proposed techniques such as multi-scale feature learning [15,16], scale normalization [17,18], or introducing super-resolution networks [19,20] to address the challenges of small object detection. However, these deep learning models require a large number of labeled data for training, while only a limited number of labeled data is available in practical applications. Meanwhile, manual labeling is inherently expensive, time-consuming, and especially challenging and error-prone for small defects. To ease the effort of acquiring a large number of labels, semi-supervised learning (SSL) is a natural fit as SSL offers a promising paradigm that leverages unlabeled data to improve model performance [21]. However, much of recent progress in SSL has focused on image classification tasks, such as [21–24]. In our case, it is vital to obtain accurate, pixel-level labels to understand the number of dies impacted by the defect. Thus, we formalize the task of detecting small defects as a segmentation problem.

There have been several approaches proposed for semi-supervised semantic segmentation [25–28]. However, they are mostly consistency-regularization-based methods, which enforce the network output to be invariant to the input perturbations [25–27]. Though these methods have reported encouraging results, they become less effective for small defects as the information contained in the few pixels of a small defect can be lost due to perturbations of the input.

Pseudo labeling [29] is another SSL strategy to utilize the limited labeled data to predict labels for unlabeled data, where the model is encouraged to produce high-confidence predictions. While it is a simple heuristic and does not require augmentations, some prior works suggest that pseudo labeling alone is not competitive as other SSL methods [30]. The reason is due to poor network calibration, or threshold setting used in the conventional pseudo labeling methods usually resulting in many incorrect pseudo labels, which in turn leads to a poor generalization of a model [24]. In this work, we use incorrect pseudo labels and noisy pseudo labels or noisy predictions in pseudo labeling interchangeably. Several works propose to combine pseudo labeling with consistency training, such as [28,31]. However, these proposed methods are primarily for medium or large objects and are often unsuitable for small defects. For example, PseudoSeg [28] uses multiple predictions obtained from class activation map (CAM) [32] to calibrate pseudo labels. However, CAM is ineffective in locating the target regions of small defects due to too few pixels contained in small defects. In other words, CAM cannot provide multiple reliable predictions for pseudo labels calibration, thus making PseudoSeg [28] less effective in detecting small defects.

To address these challenges and limitations, we propose a distribution-aware pseudo labeling method (DAP-SDD) to detect small defects precisely while effectively using limited labeled data. To the best of our knowledge, there is no existing method based on distribution-aware pseudo labeling for a semantic segmentation model. Our key contributions are summarized as follows:

- We propose a distribution-aware pseudo labeling method for small defect detection (DAP-SDD) that maximizes the use of the limited number of labels available. Bootstrapping is applied on the limited available labels to obtain an approximate distribution of the complete labels, effectively guiding the pseudo labeling propagation.
- We utilize the approximate distribution in conjunction with t-distribution confidence interval and adaptive training strategies in our proposed threshold setting method, thereby dynamically generating more pseudo labels with high confidence while reducing confirmation bias.
- We conduct extensive experiments on various datasets to validate the proposed method. The evaluation results demonstrate the effectiveness of our proposed approach that outperforms the state-of-the-art techniques.

2. Related Work

Small Object Detection. In recent years, numerous deep learning models such as [3–6] have been proposed and demonstrated impressive progress on detection performance. However, these models focus on tuning for detecting general objects, mostly of medium or large size, thus suffering from a performance bottleneck for small object detection. There are several approaches proposed to address the challenges of detecting small objects. For example, Kisantal et al. [12] applied data augmentation techniques to increase the number of small objects to improve the detection performance of the model. The authors of [15,16,33] used a multi-scale feature pyramid and deconvolution layers to improve detection performance on small and large objects. SNIP [17] proposed scale normalization and [34] used a dilated convolution network to improve the performance of detecting small objects. These approaches aimed to mitigate the imbalanced distribution of small objects from conventional object sizes. However, they still require a substantial amount of labeled data for training, which is not viable when limited labeled data are available.

Semi-supervised Semantic Segmentation. There are two common strategies used in SSL: consistency regularization and entropy minimization. In consistency regularization-based methods, the prediction is enforced to be consistent when using data augmentation for input images [25], perturbation for embedding features [26], or different networks [35]. While these methods reported impressive detection performance, they become less effective for small defects because of a limited number of pixels in small objects, which could be ignored or even lost when the input or embedding features are perturbed in consistency regularization-based methods. In this way, the model fails to learn key features to distinguish small defects from the background or other categories. On the other hand, entropy minimization encourages a model to predict low-entropy outputs for unlabeled data. Pseudo labeling [29] is one of the implicit entropy minimization methods [36]. Pseudo labeling is usually used with a high confidence threshold setting to reduce the introduction of noisy predictions. With more high confidence information incorporated, the model would learn to minimize output entropy better. However, due to suboptimal threshold setting mechanisms in the conventional pseudo labeling methods [24], some prior works suggest pseudo labeling on its own is not competitive as other SSL methods [30]. Ref. [28] combines consistency regularization with pseudo labeling to improve model performance. However, it still requires consistency regularization, which is ineffectual for small defects. Our design of pseudo labels is inspired by recent SSL-based image classification works [21–23], which incorporated distribution alignment to generate high confidence pseudo labels for unlabeled data. While these approaches require data augmentation to generate multiple class distributions for distribution alignment or comparison, our method does not require data augmentation during pseudo labeling. In addition to these two main categories of SSL methods, several GAN-based models are proposed. For example, Souly et al. [37] generates additional training data via GAN to alleviate the lack of labeled data. Hung et al. [38], on the other hand, uses an adversarial network to learn a discriminator between the ground truth and the prediction to generate a confidence map. Unlike GAN-based models, which require adversarial networks to generate additional data, our method directly generates labels via proper threshold setting without introducing extra data.

3. Methodology

Figure 2 depicts an overview of our proposed method: distribution-aware pseudo labeling for small defect detection (DAP-SDD).

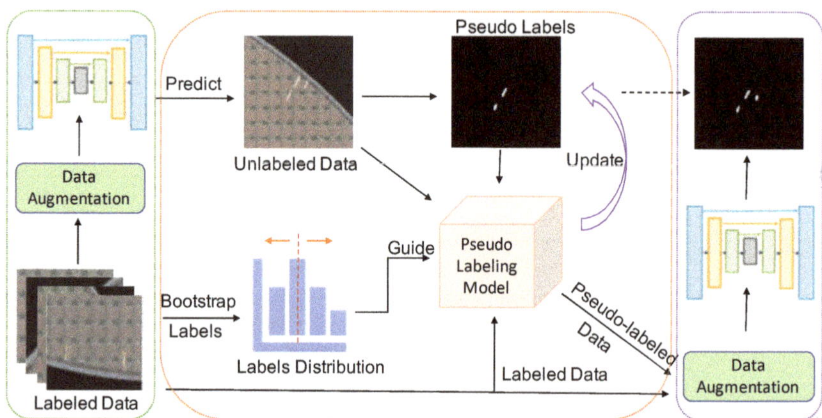

Figure 2. An overview of our proposed distribution-aware pseudo labeling for small defect detection (DAP-SDD). We first use data augmentation techniques to leverage the limited labeled data for training in Step 1 (green dash box). Then, we use the trained model to generate initial pseudo labels for unlabeled data. We also apply bootstrapping for the limited labels to obtain approximate distribution with statistics such as that of the whole labeled dataset. Then, we use it to guide the threshold setting during pseudo labeling propagation in Step 2 (orange dash box). To achieve better detection performance from our model, we update pseudo labels for unlabeled data iteratively. Once the detection performance remains or starts to degrade, we apply the warm restart and mixup augmentation for both labeled data and pseudo labeled data in Step 3 (purple dash box). This step is to overcome confirmation bias [39] and overfitting, thereby improving model performance further.

3.1. Leverage Labeled Data

Data augmentation can usually help improve detection performance, and there are several commonly used augmentation techniques we could employ, such as random crop, rotation, horizontal flip, color jittering, mixup, etc. [28,40]. However, these commonly-used techniques become incompetent in improving model performance when limited labels are available, e.g., less than 5% of the fully-labeled dataset. Inspired by augmentation techniques proposed in [12,41], we rotate images by 0, 90, 180, 270 degrees to quadruple labeled data. Unlike augmenting images by directly copy-pasting multiple times as in [12], our variants of original data not only enrich the labeled data but also can help prevent the model from becoming biased when the amount of pseudo labeled data increases during pseudo label propagation.

3.2. Distribution-Aware Pseudo Labeling

In pseudo labeling methods, one common way is to convert model predictions to hard pseudo labels directly. To illustrate this, let $D_l = \{x_l^{(i)}, y_l^{(i)}\}$ be a labeled dataset and $D_u = \{x_u^{(i)}\}$ be a unlabeled dataset. We first train a model f_θ on D_l and use the trained model to infer on D_u. Let us further denote $p(x_u^{(i)})$ as the prediction of unlabeled sample $x_u^{(i)}$, then the pseudo label for $x_u^{(i)}$ can be denoted as:

$$\tilde{y}_u^{(i)} = \mathbb{1}[p(x_u^{(i)}) > \gamma], \tag{1}$$

where $\gamma \in (0,1)$ is a threshold to generate pseudo labels. Note that, for a semantic segmentation model such as ours, $p(x_u^{(i)})$ is a probability map and $\tilde{y}_u^{(i)}$ is a binary mask with pseudo labels. As Equation (1) shows, threshold setting is critical to generate reliable pseudo labels. However, determining an optimal threshold is difficult, and a sub-optimal threshold value can introduce many incorrect pseudo labels, which degrades model performance. Therefore, we propose a novel threshold setting method, which can generate more pseudo labels with high confidence without bringing many noisy predictions.

3.2.1. Bootstrap Labels

Assuming the distribution of limited labeled data approximates that of fully labeled data, we first apply bootstrapping, a resampling technique that estimates summary statistics (e.g., mean and standard deviation) on a population by randomly sampling a dataset with replacement. The metric we employed in bootstrapping is *label_pixel_ratio*, which is denoted as:

$$label\_pixel\_ratio = \frac{label\_pixels}{image\_pixels}, \tag{2}$$

where *label_pixels* is the number of pixels of one label, and the *image_pixels* is the number of pixels of the image in which the label locates. For example, if the number of pixels of one label is 256, and the image size is 2048 × 2048, then the *label_pixel_ratio* for this label is $256/(2048 \times 2048) = 0.00006104$.

3.2.2. Distribution-Aware Pseudo Label Threshold Setting

Once we obtain the mean of *label_pixel_ratio* (μ) in the previous step, we calculate the number (k) of pixels of predictions on D_l ($p(D_l)$), and the top k of sorted $p(D_l)$ are pixels for labels. In other words, the threshold for D_l is the k-th value in $p(D_l)$, which we use to set the threshold for unlabeled data D_u as well. This mechanism works as both labeled and unlabeled data are supposed to be sampled from the same distribution of fully labeled data and share the same mean of label distribution. We also use the same trained model to infer on them. The k at a specific iteration n with the predictions of $p_n(D_l)$ is given by:

$$k_{n,base} = \lfloor \mathcal{N}(p_n(D_l)) * \mu \rceil, \tag{3}$$

where $\mathcal{N}(p_n(D_l))$ is used to obtain the total number of pixels in $p_n(D_l)$. The raw outcome from this equation is a real number, so we round it to the nearest integer to obtain $k_{n,base}$. Then, the corresponding $k_{n,base}$-th value in $p_n(D_l)$ can be used for threshold setting.

Using Equation (3), we can set a quite reasonable initial threshold as the calculation utilizes the mean of estimated label distribution. However, as the pseudo labeling model is encouraged to produce more high-confidence (i.e., low-entropy) predictions as training continues, this method alone may suffer from an insufficient number of proposed pseudo labels. To incorporate more pseudo labels with high-confidence predictions while reducing the possibility of introducing noisy predictions, we use a confidence interval and gradually increase it. An increasing confidence interval allows incorporating a higher number of confident predictions as high-confidence pseudo labels. Specifically, we use the t-distribution to find a given $100(1-\alpha)\%$ confidence interval (CI), which can be obtained via:

$$CI = \mu \pm t_{\alpha/2,m-1}\frac{s}{\sqrt{m}}, \tag{4}$$

where μ and s are the estimated mean and sample standard deviation of *label_pixel_ratio*, respectively. m represents the number of labels and t is a critical value in t-distribution table to obtain $P(T \leq t) = 1 - \alpha/2$ at the degrees freedom of $m-1$. The lower bound of the confidence interval (CI_{lower}) is $\mu - t_{\alpha/2,m-1}\frac{s}{\sqrt{m}}$, whereas the upper bound (CI_{upper}) is $\mu + t_{\alpha/2,m-1}\frac{s}{\sqrt{m}}$. We use the t-distribution in our proposed method because of the lack of labeled samples available. In such a case, the estimated standard deviation tends to be farther from the real standard deviation, and t-distribution fits better than the normal distribution. We also present the comparison results of them in the later section of ablation studies. Once we obtain the confidence interval, we can map them to find the lower and upper bound of k via Equation (3) by replacing the μ with CI_{lower} or CI_{upper}. Then, we can use the $k_{n,ci}$-th value of $p_n(D_l)$, with a given $100(1-\alpha_n)\%$ confidence level to obtain the threshold γ_n at a specific iteration n:

$$\gamma_n = \mathcal{K}(p_n(D_l), \lfloor \mathcal{N}(p_n(D_l)) * t_{\alpha_n/2,m-1}\frac{2s}{\sqrt{m}} * \nu_n \rceil), \tag{5}$$

where $\mathcal{K}(p_n, k)$ is a function to find k-th value in p_n and ν_n is an adjustment factor used to slow down or speed up propagation during training.

In addition to using the t-distribution to calculate the confidence interval for setting thresholds, we also employ another intuitive method for selecting high confidence pseudo labels. Specifically, we find the threshold that produces the best performance on labeled data. We then use that threshold to generate initial pseudo labels for unlabeled data. To illustrate this, let $P_{l,0}$ denote the precision obtained from the labeled data, which can be considered as a confidence level for pseudo labels since the precision indicates how many predictions out of all predictions are true small defects. We can increase the confidence level with a moving step τ as the training goes on. Along with the $k_{n,base}$ via Equation (3), we can obtain the threshold γ_n at a specific iteration n by using:

$$\gamma_n = \mathcal{K}(p_n(D_l), \lfloor \mathcal{N}(p_n(D_l)) * \mu * (P_{l,0} + \nu_n * \tau) \rceil). \tag{6}$$

Overall, the method utilizing t-distribution confidence interval Equation (5) performs better than the intuitive method Equation (6), and their comparison results are presented in the later section of ablation studies.

3.2.3. Training Strategies

During training, we adjust the moving step of pseudo labeling propagation to set threshold adaptively. To accomplish this, we keep monitoring training and use the model evaluation results (e.g., Precision, Recall, F1 score) on labeled data. For example, if the monitored results show a decrease in both F1 score and recall but an increase in precision (close to 1.0), it indicates the threshold is set too high to incorporate more confident pseudo labels. In other words, the model can speed up the propagation and set the adjustment factor ν to a bigger value so that the threshold will be set to a lower value, thereby incorporating more high-confidence pseudo labels and vice versa. Another strategy we adopt during training is a weighted moving average of thresholds. Due to the random combination of training data batch, a model may temporarily suffer a significant performance decrease in a certain iteration. A weighted moving average of thresholds can prevent such an outlier threshold from resetting the threshold value that the model has learned to ensure more stable pseudo labeling propagation.

Algorithm 1 presents the training procedure of our proposed distribution-aware pseudo labeling. First, we use the labeled data to train a model $f_{\theta,0}$, and then use the trained model to generate initial pseudo labels and obtain the initial precision $P_{l,0}$. During the iterative pseudo labeling with the maximum number of iterations N, we calculate thresholds γ_n for each iteration via Equations (5) or (6). Then, we evaluate the obtained thresholds γ_n and the moving average threshold $\gamma_{n-1,ma}$ on labeled data. The threshold that yields better evaluation results (i.e., F1 Score) is selected. Meanwhile, by comparing the evaluation results (Precision, Recall, F1 Score denoted as $P_n, R_n, F1_n$, respectively) of the current iteration with that of the previous iteration, we can obtain the adjustment factor ν_n to speed up or slow down pseudo label propagation. Moreover, we update the moving average threshold for the next iteration. Next, we use the selected threshold γ_n to generate pseudo labels for unlabeled data D_u. We then combine the pseudo labeled data D_p with labeled data D_l to retrain the model. We repeat these steps to update pseudo labels iteratively to achieve better detection performance of the model. Once the detection performance from the model reaches a certain threshold (e.g., F1 $\geq$ 0.85) but remains or decreases beyond that, the warm restart and mixup augmentation will be applied on both labeled data and pseudo labeled data to improve detection performance further.

Algorithm 1 Distribution-Aware Pseudo Labeling.

```
Train a model f_{θ,0} using labeled data D_l.
for n = 1,2,...,N do
    Obtain threshold γ_n
    P_n, R_n, F1_n, ν_n, γ_n ← E(D_l, γ_n, γ_{n−1,ma})
    γ_{n,ma} ← M(γ_n, γ_{n−1}, γ_{n−2}, α, β)
    D_{p,n} ← Pseudo label D_u using γ_n
    D̃ ← D_l ∪ D_{p,n}
    Train f_{θ,n} using D̃.
    f_θ, D_p ← f_{θ,n}, D_{p,n}
end for
return f_θ, D_p
```

3.2.4. Loss Function

During pseudo labeling propagation, the loss function $\mathcal{L}_p$ incorporates labeled and pseudo labeled data, which can be denoted as:

$$\mathcal{L}_p = -\Big(\sum_{D_l}\mathcal{L}(y_l, \hat{y}_l) + \eta\sum_{D_u}\mathcal{L}(y_p, \hat{y}_p)\Big), \tag{7}$$

where y_l is the ground truth labels, and y_p represents pseudo labels. $\hat{y}_l$ and $\hat{y}_p$ denote the predictions of labeled data and unlabeled data, respectively. $\mathcal{L}$ represents the cross-entropy loss function. As the pseudo labeling progresses, the amount of pseudo labeled data will increase accordingly. To avoid the model increasingly favoring pseudo labeled data over the original labeled data, we add a weight $\eta \in (0,1)$ to adjust the impact from pseudo labels. In practice, we can achieve this by repeatedly sampling or using the similar augmentation techniques described in Section 3.1.

During the training process using mixup augmentation, we define a mixup loss function $\mathcal{L}_m$, which is given by:

$$\mathcal{L}_m = -\sum_{i=1}^{N}\Big(\lambda\mathcal{L}(y_a^{(i)}, \hat{y}_a^{(i)}) + (1-\lambda)\mathcal{L}(y_b^{(i)}, \hat{y}_b^{(i)})\Big), \tag{8}$$

y_a and y_b are the original labels of the input images, and $\hat{y}_a$ and $\hat{y}_b$ are corresponding predictions. N is the number of samples used for training. In the first step of leveraging labeled data, N only includes the number of labeled data, while in the last step, both the labeled and pseudo labeled data will be included. $\lambda \in [0,1]$ is used in mixup augmentation for constructing virtual inputs and outputs [40]. Specifically, the mixup uses the following rules to create virtual training examples:

$$\tilde{x} = \lambda \times x_a + (1-\lambda) \times x_b$$

$$\tilde{y} = \lambda \times y_a + (1-\lambda) \times y_b,$$

where (x_a, y_a) and (x_b, y_b) are two original inputs drawn at random from training batch, $\lambda \in [0,1]$ and the $\tilde{x}$ and $\tilde{y}$ are constructed input and corresponding output.

4. Results and Discussion

4.1. Datasets

We use an in-house dataset from the wafer inspection system (WIS) in our evaluation. This dataset contains two types of small defects on wafer images: edge void and arc-like defects. Each of them has 213 images: 173 for training and 40 for test. There are 618 labels for edge void and 406 labels for arc-like, weak labels from the current system tool and verified predictions from a trained model. The image size is 2048 × 2048, and we crop it into 512 × 512 patches to fit into GPU memory.

We also evaluate our method on two public datasets: industrial optical inspection dataset of DAGM 2007 [14] and tiny defect detection dataset for PCB [42]. We use Class 8 and Class 9 of DAGM as they fit into a small defect category (denoted as Dagm-1 and Dagm-2). We split each class into two sets of 150 defective images in gray-scale for training and testing. The image size is 512 × 512 in DAGM. The defects are labeled as ellipses, and each image has one labeled defect.

On the other hand, the PCB dataset includes six types of tiny defects (missing hole, mouse bite, open circuit, short, spur, and spurious copper), and each image may have multiple defects. PCB contains 693 images with defects: 522 and 101 images for training and test, respectively. The total number of defects is 2953. There are different sizes of PCB images, and the average pixel size of an image is 2777 × 2138. We also crop it into 512 × 512 patches for training.

4.2. Evaluation Metrics

Intersection over Prediction (IoP). In this work, instead of using IoU (intersection over union), we adopt IoP (intersection over prediction) [43] to overcome the issue shown in Figure 1, where one weak label may contain multiple small defects or cover more area than the true defect area. IoP is defined as the intersection area between ground truth and prediction divided by the area of prediction. If the IoP of a prediction for a small defect is greater than a given threshold (0.5 in this work), we count it as a true positive; otherwise, we count it as a false positive. If one weak label contains multiple true positive predictions, we only count it as one true positive.

Average Precision (AP), F1 Score. We use AP (average precision) and F1 Score to evaluate the performance of small defect detection.

4.3. Experimental Settings and Parameters

In this work, we adopt a commonly-used segmentation model U-Net [8] in our proposed method, which has been proven to be effective in medical image segmentation tasks, such as detecting microcalcifications in mammograms [43,44]. Moreover, U-Net has a relatively small size of model parameters, which is favorable in practical use. U-Net consists of three downsampling blocks and three upsampling blocks with skip connections. Each block has two convolution layers, and each of them is followed by batch normalization and ReLU. As our proposed pseudo labeling strategy is not confined to a specific deep learning model, it can be easily implemented in other deep neural networks. We will extend our proposed techniques to other segmentation models in future work.

We use the Adam optimizer in the model training. The initial learning rate is set to 1×10^{-3} and gradually decreases during training. The adjustment factor ν_n is set to 1.1 to speed up pseudo label propagation or set to 0.9 to slow down the propagation. The moving average weights $[\alpha, \beta, (1-\alpha-\beta)]$ are set to $[0.5, 0.3, 0.2]$ for the current iteration threshold γ_n and thresholds of previous two iterations $\gamma_{n-1}, \gamma_{n-2}$, respectively. The t-distribution confidence interval ranges from 0.5 to 0.995 with a moving step of 0.005.

4.4. Experiment Results

We first evaluate our proposed method on two different types of small defects on wafer images (the WIS dataset). Figure 3 demonstrates the improvements brought by our method over the supervised baseline. Overall, our proposed method can achieve above

0.93 of AP for a different amount of labeled data available and obtain comparable results as a fully-labeled (100%) dataset even when the labeled data ratio is 2% (four labeled images). However, the detection performance of the supervised method decreases dramatically when the labeled data size is limited. For instance, the AP reduces to below 0.6 when 2% of labeled data is available.

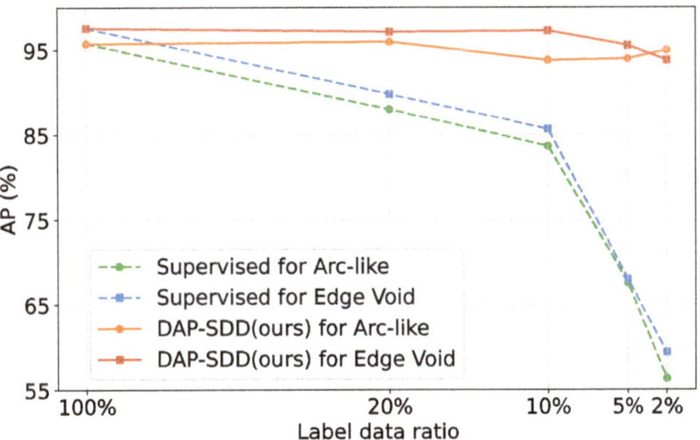

Figure 3. Improvement over the supervised baseline on two small defects in the WIS dataset.

Figure 4 demonstrates the improvements by our method (solid lines) over the supervised baseline (dash lines) on the DAGM and PCB datasets. Similar to the results of the WIS dataset, our proposed method can achieve comparative results of fully-labeled (100%) when the labeled data ratio is 10% on different small defects in DAGM and PCB datasets. The average precision (AP) by our method remains above 0.9 when only 5% of labeled data are available while the AP by the supervised model degrades dramatically. Note that in Figure 4, the values of PCB-average represent the average AP of six different types of defects in the PCB datasets.

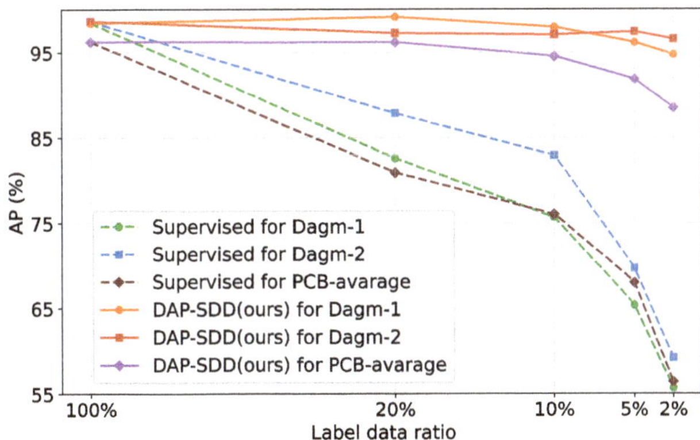

Figure 4. Improvements over the supervised baseline on small defects in the DAGM and PCB datasets.

We then compare our method with state-of-the-art semi-supervised semantic segmentation methods. Table 1 shows the comparison results on the Edge Void defect dataset with 10% of labeled data. CCT [26] is a consistency regularization-based method. As shown in Table 1, the CCT alone fails to recognize and locate small defects. CCT also provides a way of training with offline pseudo labels. So, we use the pseudo labels generated from the first step of our method. As we can see, CCT+Pseudo improves the detection performance as more pseudo labeled data are incorporated. However, the initial pseudo labels might contain incorrect labels, which are not updated iteratively in CCT. Therefore, CCT+Pseudo still presents relatively low detection performance. AdvSemSeg [38], however, uses an adversarial network for semi-supervised semantic segmentation. The experiment results show that AdvSemSeg performs better than CCT, which indicates adversarial network can be a potential direction for improving small defect detection. However, due to limited ground truth labels, AdvSemSeg does not perform well as reported in [38]. In self-training, we exclude the labeled data and only use the initial pseudo labels as the supervisory signals for unlabeled data. During self-training, instead of setting pseudo labels based on pixel confidence score higher than 0.5 as in [45], we adopt the same threshold setting strategies as our method to generate pseudo labels for self-taught training effectively. As shown in Table 1, self-training shows significantly better AP and F1 scores than CCT and AdvSegSeg. Overall, DAP-SDD achieves the highest AP and F1 scores. We attribute this to the fact that ours also incorporates labeled data that contain useful prior knowledge.

Table 1. Comparison with state-of-the-art methods on the WIS dataset with 10% of labeled data.

Method	Edge void		Arc-like	
	AP (%)	F1 (%)	AP (%)	F1 (%)
CCT (Ouali et al.)	-	-	-	-
CCT+Pseudo (Ouali et al.)	70.75	69.80	71.09	72.13
AdvSemSeg (Hung et al.)	76.61	76.42	76.98	79.51
Self-training (Zoph et al.)	89.29	88.14	85.34	86.89
DAP-SDD (Ours)	**97.29**	**96.64**	**94.99**	**91.38**

In Figure 5, we present examples of predicted labels for small defects generated by different methods. We can observe that all the evaluated models can generate labels for relatively large defects, as shown in the first row of edge void defects and the third row of arc-like defects. Compared to CCT+Pseudo or AdvSemSeg, which generate incomplete labels or overfull labels, self-training and our proposed method obtain more accurate labels. However, for the significantly tiny or dim defects, such as ones shown in the second row and fourth row, most of these models suffer from missing detection while our method can still detect them. Overall, our proposed method performs best regardless of the different sizes of small defects.

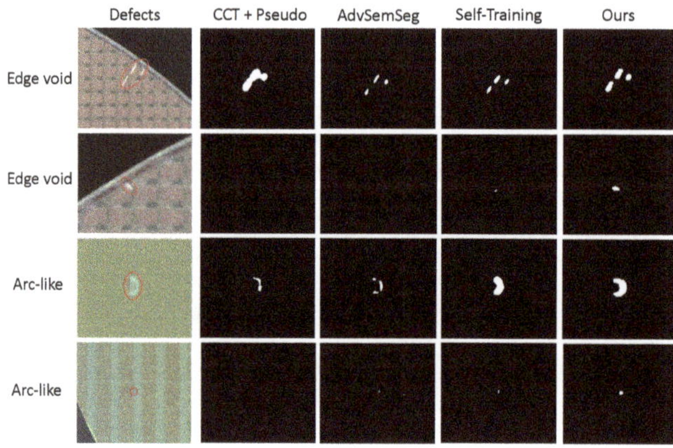

Figure 5. Examples of predicted labels using different methods for edge void and arc-like defects (marked in red) in the WIS dataset. From left to right, columns are defects, segmentation results using CCT+Pseudo, AdvSemSeg, self-training, and DAP-SDD (ours), respectively.

The prediction results and comparison results with state-of-the-art methods on the DAGM and PCB datasets are shown in Tables 2 and 3.

Table 2. Evaluation results (AP, %) on public datasets (DAGM, PCB) when different amounts of labeled data are available. Total data amount (100%): DAGM (552), DAGM (150).

Data Amount	DAGM		PCB						
	Dagm-1	Dagm-2	Missing Hole	Mouse Bite	Open Circuit	Short	Spur	Spurious Copper	Average
100%	98.46	98.65	98.75	96.03	95.34	98.57	96.38	92.46	96.26
20%	99.14	97.26	98.54	94.71	96.45	94.03	97.19	96.27	96.20
10%	97.96	97.09	97.35	93.98	91.75	92.77	95.28	96.19	94.55
5%	96.15	97.41	97.13	89.10	86.07	89.92	91.95	96.98	91.86
2%	94.74	96.54	95.67	87.71	83.07	85.79	88.32	90.38	88.49

Table 3. Comparison with state-of-the-art methods on public datasets (DAGM, PCB), evaluation metric: AP (%).

Data Amount	DAGM		PCB						
	Dagm-1	Dagm-2	Missing Hole	Mouse Bite	Open Circuit	Short	Spur	Spurious Copper	Average
CCT [26]	69.55	66.07	63.81	57.84	55.72	56.39	63.61	63.67	62.08
CCT+Pseudo [26]	83.15	82.79	79.37	73.32	70.18	72.46	79.40	76.99	75.29
AdvSemSeg [38]	84.62	85.68	84.28	80.71	80.44	82.10	83.45	83.75	82.46
Self-training [45]	92.59	91.18	88.15	85.84	86.83	85.67	87.92	87.41	86.97
DAP-SDD (Ours)	**97.96**	**97.09**	**97.35**	**93.98**	**91.75**	**92.77**	**95.28**	**96.19**	**94.55**

Moreover, we present examples of predicted labels for small defects in the DAGM (Figure 6) and PCB datasets (Figure 7) generated by different methods. As the results have shown, our proposed method consistently outperforms the state-of-the-art semi-supervised segmentation models on various datasets with various types of defects.

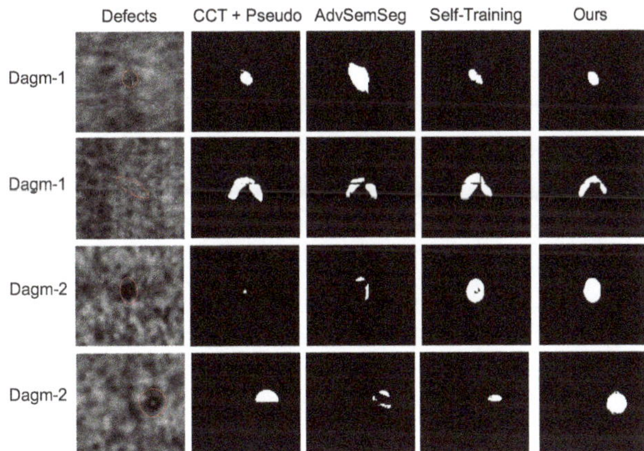

Figure 6. Examples of predicted labels using different methods on the DAGM dataset. The top two rows show results for Dagm-1, while the bottom two rows show results for Dagm-2. The first column shows defect images with original weak labels (marked in red color), and the remaining columns are segmentation results using CCT+Pseudo, AdvSemSeg, self-training, and DAP-SDD (ours), respectively.

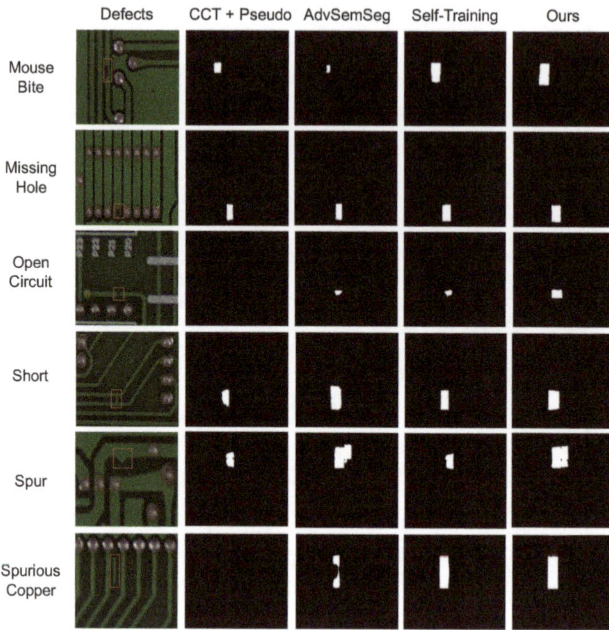

Figure 7. Examples of predicted labels using different methods on the PCB dataset. From top to bottom, each row represents six types of defects in the PCB dataset: mouse bite, missing hole, open circuit, short, spur, and spurious copper. The first column shows defect images with original weak labels (marked in red color), and the remaining columns are segmentation results using CCT+Pseudo, AdvSemSeg, self-training, and DAP-SDD (ours), respectively.

4.5. Ablation Studies

Contribution of components for performance improvement. Figure 8 demonstrates how different components in our proposed method contribute to detection performance on both in-house and public datasets. For a fair comparison, we use the same data augmentations in the supervised baseline and ours. Therefore, the results of the first step using only labeled data are also supervised baseline. As shown in Figure 8, for the WIS dataset (solid bars), the model using 20% of labeled data can achieve around 88% of AP, which is still lower than our target (our real-world applications typically require AP of 90% or higher). When we have 2% of labeled data available, the AP value decreases to 56%. In Step 2, utilizing the proposed distribution-aware pseudo labeling method significantly improved the detection performance for all cases, and cases with fewer labeled data benefit more. For example, AP is improved from 56% to 92% when using 2% of labeled data. The results demonstrate that our proposed method can effectively leverage the information from massive unlabeled data to improve detection performance. In the final step, warm restart and mixup are employed to improve performance further. We obtain similar results on public datasets shown in Figure 8 (bars with patterns). Overall, the proposed distribution-aware pseudo labeling contributes most significantly to the detection performance and the data augmentations we adopted in DAP-SDD are effective in enhancing the model's performance and robustness.

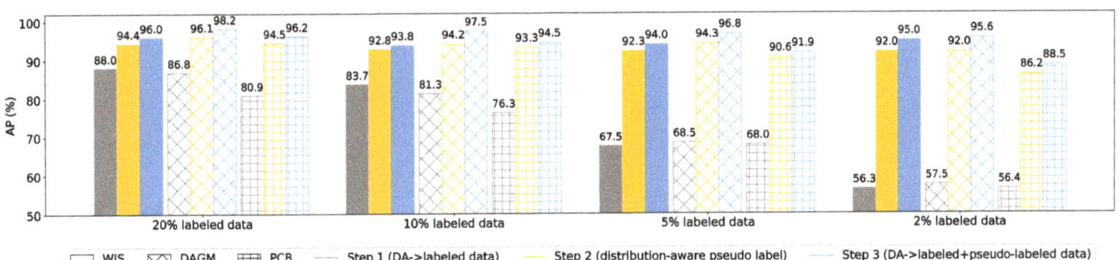

Figure 8. Ablation studies on different factors that contribute to performance improvement.

Compare with more baselines. In our proposed DAP-SDD, we assume the distribution of proposed labels approximates the distribution of ground truth labels as training proceeds. We use the Kullback–Leibler (KL) divergence to evaluate the differences in the distribution of proposed labels compared with ground truth labels during training, which is shown in Figure 9. The KL divergence is a commonly-used measurement for evaluating how one probability distribution differs from the other reference distribution. We can observe that: (a) t-dist vs. normal-dist: t-distribution (t-dist) performs better than a normal distribution (normal-dist) because t-dist has heavier tails. Thus it is more suitable for estimating the confidence interval (CI) when the sample size is limited as in our cases. For a given CI range, normal-dist tends to incorporate more predictions than t-dist, which in turn brings 'too many' noisy predictions for pseudo labels. As a result, the accumulated noisy impact overwhelms that of original limited labels as training proceeds. (b) Adaptive vs. fixed threshold: adaptive thresholding that combines Equation (3) and Equation (5) can keep the model learning more useful information during training and outperforms the fixed threshold obtained via Equation (3). (c) Equation (5) vs. Equation (6): Equation (6) is more conservative in incorporating confident predictions than Equation (5) when using the same moving step (0.005), and it requires more training epochs to reach the equivalent results as Equation (5). (d) with vs. without ma: compared with the baseline without moving average (ma) threshold, our method incorporates ma, which helps prevent outlier thresholds (e.g., epochs 23 and 65) from resetting what the model has learned. In addition, the corresponding detection performance and KL divergence at the same training epoch (100th) of these baselines are shown in Table 4. As we can observe, DAP-SDD using the t-distribution confidence interval and with moving average achieves the best detection performance while having the smallest KL divergence.

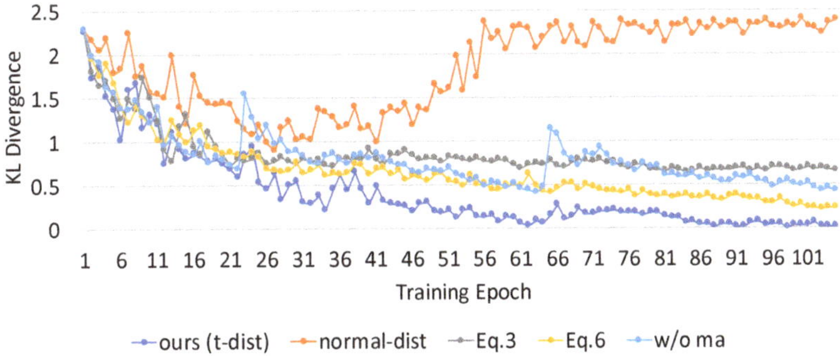

Figure 9. KL divergence curves of various baselines.

Table 4. Comparison of detection performance (AP, %) and KL divergence (same training epochs 100) on various baselines.

Baseline Method	AP (%)	KL divergence
Normal distribution	66.52	2.3772
Fixed threshold Equation (3)	81.63	0.6528
DAP-SDD via Equation (6)	94.34	0.2288
DAP-SDD w/o ma	86.98	0.4181
DAP-SDD (t-dist, Equation (5), ma)	**97.29**	**0.0127**

5. Conclusions

In this work, we propose a distribution-aware pseudo labeling for small defect detection (DAP-SDD) when limited labeled data are available. We first applied bootstrapping for the available labeled data to approximate the distribution of the whole labeled dataset. Then, we used it to guide pseudo label propagation. Our proposed method incorporates t-distribution confidence interval and adaptive training strategies, and thus can effectively generate more pseudo labels with high confidence while reducing confirmation bias. The extensive experimental evaluation on various datasets with various types of defects has demonstrated that our proposed DAP-SDD consistently outperforms the state-of-the-art techniques with above 0.9 of average precision and up to 0.99. Our in-depth analysis of the ablation studies clearly shows how each component employed in our approach effectively utilizes the limited labeled data.

Author Contributions: Conceptualization, X.Z. (Xiaoyan Zhuo); methodology, X.Z. (Xiaoyan Zhuo) and W.R.; evaluation, X.Z. (Xiaoyan Zhuo), W.R., and X.Z. (Xiaoqian Zhang); formal analysis, X.Z. (Xiaoyan Zhuo) and W.R.; writing—original draft preparation, X.Z. (Xiaoyan Zhuo); writing—review and editing, X.Z. (Xiaoyan Zhuo), W.R., and S.W.S.; supervision, T.D. and S.W.S. All authors have read and agreed to the published version of the manuscript.

Institutional Review Board Statement: Not applicable.

Informed Consent Statement: Not applicable.

Data Availability Statement: DAGM dataset: https://hci.iwr.uni-heidelberg.de/content/weakly-supervised-learning-industrial-optical-inspection (accessed on 7 April 2022). PCB dataset: http://robotics.pkusz.edu.cn/resources/dataset/ (accessed on 7 April 2022).

Acknowledgments: The Titan X Pascal used for this research was donated by the NVIDIA Corporation. The authors acknowledge the MIT SuperCloud and Lincoln Laboratory Supercomputing Center for providing (HPC, database, consultation) resources that have contributed to the research results reported within this paper.

Conflicts of Interest: The authors declare no conflict of interest.

References

1. Shankar, N.; Zhong, Z. Defect detection on semiconductor wafer surfaces. *Microelectron. Eng.* **2005**, *77*, 337–346. [CrossRef]
2. Huang, S.H.; Pan, Y.C. Automated visual inspection in the semiconductor industry: A survey. *Comput. Ind.* **2015**, *66*, 1–10. [CrossRef]
3. Ren, S.; He, K.; Girshick, R.; Sun, J. Faster R-CNN: Towards Real-Time Object Detection with Region Proposal Networks. *Adv. Neural Inf. Process. Syst.* **2015**, *28*, 91–99. [CrossRef] [PubMed]
4. Redmon, J.; Farhadi, A. YOLOv3: An Incremental Improvement. *arXiv* **2018**, arXiv:1804.02767.
5. He, K.; Gkioxari, G.; Dollár, P.; Girshick, R. Mask R-CNN. In Proceedings of the 2017 IEEE International Conference on Computer Vision (ICCV), Venice, Italy, 22–29 October 2017; pp. 2980–2988. [CrossRef]
6. Lin, T.; Goyal, P.; Girshick, R.; He, K.; Dollár, P. Focal Loss for Dense Object Detection. In Proceedings of the 2017 IEEE International Conference on Computer Vision (ICCV), Venice, Italy, 22–29 October 2017; pp. 2999–3007. [CrossRef]
7. Long, J.; Shelhamer, E.; Darrell, T. Fully convolutional networks for semantic segmentation. In Proceedings of the 2015 IEEE Conference on Computer Vision and Pattern Recognition (CVPR), Boston, MA, USA, 7–12 June 2015; pp. 3431–3440. [CrossRef]
8. Ronneberger, O.; Fischer, P.; Brox, T. U-Net: Convolutional Networks for Biomedical Image Segmentation. In *International Conference on Medical Image Computing and Computer-Assisted Intervention*; Springer: Cham, Switzerland, 2015.

9. Chen, L.C.; Zhu, Y.; Papandreou, G.; Schroff, F.; Adam, H. Encoder-Decoder with Atrous Separable Convolution for Semantic Image Segmentation. In Proceedings of the European Conference on Computer Vision (ECCV), Munich, Germany, 8–14 September 2018.
10. Tong, K.; Wu, Y.; Zhou, F. Recent advances in small object detection based on deep learning: A review. *Image Vis. Comput.* **2020**, *97*, 103910. [CrossRef]
11. Fu, K.; Li, J.; Ma, L.; Mu, K.; Tian, Y. Intrinsic Relationship Reasoning for Small Object Detection. *arXiv* **2020**, arXiv:2009.00833.
12. Kisantal, M.; Wojna, Z.; Murawski, J.; Naruniec, J.; Cho, K. Augmentation for small object detection. *arXiv* **2019**, arXiv:1902.07296.
13. Nguyen, N.D.; Do, T.; Ngo, T.D.; Le, D.D. An Evaluation of Deep Learning Methods for Small Object Detection. *J. Electr. Comput. Eng.* **2020**, *2020*, 3189691. [CrossRef]
14. Heidelberg Collaboratory for Image Processing (HCI). *DAGM 2007 Competition Dataset: Industrial Optical Inspection Dataset.* Heidelberg Collaboratory for Image Processing, Heidelberg University: Heidelberg, Germany. Available online: https://hci.iwr.uni-heidelberg.de/content/weakly-supervised-learning-industrial-optical-inspection (accessed on 7 April 2022).
15. Fu, C.Y.; Liu, W.; Ranga, A.; Tyagi, A.; Berg, A.C. DSSD: Deconvolutional Single Shot Detector. *arXiv* **2017**, arXiv:1701.06659.
16. Cao, G.; Xie, X.; Yang, W.; Liao, Q.; Shi, G.; Wu, J. Feature-fused SSD: fast detection for small objects. Ninth International Conference on Graphic and Image Processing (ICGIP 2017), Qingdao, China, 14–16 October 2017; Volume 10615, p. 106151E.
17. Singh, B.; Davis, L.S. An Analysis of Scale Invariance in Object Detection-SNIP. In Proceedings of the 2018 IEEE/CVF Conference on Computer Vision and Pattern Recognition (CVPR), Salt Lake City, UT, USA, 18–23 June 2018. [CrossRef]
18. Singh, B.; Najibi, M.; Davis, L.S. SNIPER: Efficient Multi-Scale Training. In Proceedings of the 32nd Conference on Neural Information Processing Systems (NeurIPS 2018), Montréal, QC, Canada, 3–8 December 2018. pp. 9333–9343.
19. Bai, Y.; Zhang, Y.; Ding, M.; Ghanem, B. Finding Tiny Faces in the Wild with Generative Adversarial Network. In Proceedings of the 2018 IEEE/CVF Conference on Computer Vision and Pattern Recognition (CVPR), Salt Lake City, UT, USA, 18–23 June 2018; pp. 21–30. [CrossRef]
20. Bai, Y.; Zhang, Y.; Ding, M.; Ghanem, B. SOD-MTGAN: Small Object Detection via Multi-Task Generative Adversarial Network. In *Computer Vision–ECCV 2018*; Ferrari, V., Hebert, M., Sminchisescu, C., Weiss, Y., Eds.; Springer: Cham, Switzerland, 2018; pp. 210–226.
21. Berthelot, D.; Carlini, N.; Goodfellow, I.; Papernot, N.; Oliver, A.; Raffel, C.A. MixMatch: A Holistic Approach to Semi-Supervised Learning. *Adv. Neural Inf. Process. Syst.* **2019**, *32*, 5049–5059.
22. Berthelot, D.; Carlini, N.; Cubuk, E.D.; Kurakin, A.; Sohn, K.; Zhang, H.; Raffel, C. ReMixMatch: Semi-Supervised Learning with Distribution Matching and Augmentation Anchoring. *arXiv* **2020**, arXiv:1911.09785.
23. Kurakin, A.; Li, C.L.; Raffel, C.; Berthelot, D.; Cubuk, E.D.; Zhang, H.; Sohn, K.; Carlini, N.; Zhang, Z. FixMatch: Simplifying Semi-Supervised Learning with Consistency and Confidence. *Adv. Neural Inf. Process. Syst.* **2020**, *33*, 596–608.
24. Rizve, M.N.; Duarte, K.; Rawat, Y.S.; Shah, M. In Defense of Pseudo-Labeling: An Uncertainty-Aware Pseudo-label Selection Framework for Semi-Supervised Learning. *arXiv* **2021**, arXiv:2101.06329.
25. French, G.; Laine, S.; Aila, T.; Mackiewicz, M.; Finlayson, G.D. Semi-supervised semantic segmentation needs strong, varied perturbations. In Proceedings of the 31st British Machine Vision Conference 2020, BMVC 2020, Virtual Event, UK, 7–10 September 2020.
26. Ouali, Y.; Hudelot, C.; Tami, M. Semi-Supervised Semantic Segmentation With Cross-Consistency Training. In Proceedings of the 2020 IEEE/CVF Conference on Computer Vision and Pattern Recognition (CVPR), Seattle, WA, USA, 13–19 June 2020; pp. 12671–12681.
27. Verma, V.; Lamb, A.; Kannala, J.; Bengio, Y.; Lopez-Paz, D. Interpolation Consistency Training for Semi-supervised Learning. In Proceedings of the Twenty-Eighth International Joint Conference on Artificial Intelligence (IJCAI 2019), Macao, China, 10–16 August 2019. pp. 3635–3641. [CrossRef]
28. Zou, Y.; Zhang, Z.; Zhang, H.; Li, C.L.; Bian, X.; Huang, J.B.; Pfister, T. PseudoSeg: Designing Pseudo Labels for Semantic Segmentation. *arXiv* **2021**, arXiv:2010.09713.
29. Lee, D. Pseudo-Label : The Simple and Efficient Semi-Supervised Learning Method for Deep Neural Networks. In Proceedings of the 30th International Conference on Machine Learning (ICML) Workshop, Atlanta, GA, USA, 16–21 June 2013.
30. Oliver, A.; Odena, A.; Raffel, C.A.; Cubuk, E.D.; Goodfellow, I. Realistic Evaluation of Deep Semi-Supervised Learning Algorithms. In Proceedings of the 32nd Conference on Neural Information Processing Systems (NeurIPS 2018), Montréal, QC, Canada, 3–8 December 2018. pp. 3239–3250.
31. Chen, Z.; Zhang, R.; Zhang, G.; Ma, Z.; Lei, T. Digging Into Pseudo Label: A Low-Budget Approach for Semi-Supervised Semantic Segmentation. *IEEE Access* **2020**, *8*, 41830–41837. [CrossRef]
32. Zhou, B.; Khosla, A.; Lapedriza, A.; Oliva, A.; Torralba, A. Learning Deep Features for Discriminative Localization. In Proceedings of the 2016 IEEE Conference on Computer Vision and Pattern Recognition (CVPR), Las Vegas, NV, USA, 27–30 June 2016.
33. Liu, W.; Anguelov, D.; Erhan, D.; Szegedy, C.; Reed, S.; Fu, C.Y.; Berg, A.C. SSD: Single Shot MultiBox Detector. *arXiv* **2015**, arXiv:1512.02325.
34. Li, Y.; Chen, Y.; Wang, N.; Zhang, Z. Scale-Aware Trident Networks for Object Detection. In Proceedings of the 2019 IEEE/CVF International Conference on Computer Vision (ICCV), Seoul, Korea, 27 October–2 November 2019; pp. 6053–6062. [CrossRef]
35. Ke, Z.; Qiu, D.; Li, K.; Yan, Q.; Lau, R.W. Guided Collaborative Training for Pixel-wise Semi-Supervised Learning. In *European Conference on Computer Vision*; Springer: Cham, Switzerland, 2020.

36. Grandvalet, Y.; Bengio, Y. Semi-supervised Learning by Entropy Minimization. *Adv. Neural Inf. Process. Syst.* **2005**, *17*, 529–536.
37. Souly, N.; Spampinato, C.; Shah, M. Semi Supervised Semantic Segmentation Using Generative Adversarial Network. In Proceedings of the 2017 IEEE International Conference on Computer Vision (ICCV), Venice, Italy, 22–29 October 2017; pp. 5689–5697. [CrossRef]
38. Hung, W.C.; Tsai, Y.H.; Liou, Y.T.; Lin, Y.Y.; Yang, M.H. Adversarial Learning for Semi-supervised Semantic Segmentation. *arXiv* **2018**, arXiv:1802.07934.
39. Arazo, E.; Ortego, D.; Albert, P.; O'Connor, N.E.; McGuinness, K. Pseudo-Labeling and Confirmation Bias in Deep Semi-Supervised Learning. In Proceedings of the 2020 International Joint Conference on Neural Networks (IJCNN), Glasgow, UK, 19–24 July 2020; pp. 1–8. [CrossRef]
40. Zhang, H.; Cisse, M.; Dauphin, Y.N.; Lopez-Paz, D. mixup: Beyond Empirical Risk Minimization. *arXiv* **2017**, arXiv:1710.09412.
41. Gidaris, S.; Singh, P.; Komodakis, N. Unsupervised Representation Learning by Predicting Image Rotations. *arXiv* **2018**, arXiv:1803.07728.
42. Printed Circuit Board (PCB) Tiny Defects Dataset, Open Lab on Human Robot Interaction of Peking University, Beijing, China. Available online: http://robotics.pkusz.edu.cn/resources/dataset/ (accessed on 7 April 2022).
43. Cao, Z.; Yang, Z.; Zhuo, X.; Lin, R.; Wu, S.; Huang, L.; Han, M.; Zhang, Y.; Ma, J. DeepLIMa: Deep Learning Based Lesion Identification in Mammograms. In Proceedings of the 2019 IEEE/CVF International Conference on Computer Vision Workshop (ICCVW), Seoul, Korea, 27–28 October 2019; pp. 362–370. [CrossRef]
44. Zhang, F.; Luo, L.; Sun, X.; Zhou, Z.; Li, X.; Yu, Y.; Wang, Y. Cascaded Generative and Discriminative Learning for Microcalcification Detection in Breast Mammograms. In Proceedings of the 2019 IEEE/CVF Conference on Computer Vision and Pattern Recognition (CVPR), Long Beach, CA, USA, 15–20 June 2019; pp. 12570–12578. [CrossRef]
45. Zoph, B.; Ghiasi, G.; Lin, T.Y.; Cui, Y.; Liu, H.; Cubuk, E.D.; Le, Q.V. Rethinking Pre-training and Self-training. *arXiv* **2020**, arXiv:2006.06882.

MDPI

Proceeding Paper

Quantifying Bias in a Face Verification System †

Megan Frisella [1,*,‡], Pooya Khorrami [2], Jason Matterer [2], Kendra Kratkiewicz [2] and Pedro Torres-Carrasquillo [2]

1 Department of Mathematics, Brown University, Providence, RI 02912, USA
2 Lincoln Laboratory, Massachusetts Institute of Technology, Lexington, MA 02421, USA; pooya.khorrami@ll.mit.edu (P.K.); jason.matterer@ll.mit.edu (J.M.); kendra@ll.mit.edu (K.K.); ptorres@ll.mit.edu (P.T.-C.)
* Correspondence: megan_frisella@brown.edu
† Presented at the AAAI Workshop on Artificial Intelligence with Biased or Scarce Data (AIBSD), Online, 28 February 2022.
‡ Work done while author was an intern at MIT Lincoln Laboratory.

Abstract: Machine learning models perform face verification (FV) for a variety of highly consequential applications, such as biometric authentication, face identification, and surveillance. Many state-of-the-art FV systems suffer from unequal performance across demographic groups, which is commonly overlooked by evaluation measures that do not assess population-specific performance. Deployed systems with bias may result in serious harm against individuals or groups who experience underperformance. We explore several fairness definitions and metrics, attempting to quantify bias in Google's FaceNet model. In addition to statistical fairness metrics, we analyze clustered face embeddings produced by the FV model. We link well-clustered embeddings (well-defined, dense clusters) for a demographic group to biased model performance against that group. We present the intuition that FV systems underperform on protected demographic groups because they are less sensitive to differences between features within those groups, as evidenced by clustered embeddings. We show how this performance discrepancy results from a combination of representation and aggregation bias.

Keywords: face verification; bias; fairness

Citation: Frisella, M.; Khorrami, P.; Matterer, J.; Kratkiewicz, K.; Torres-Carrasquillo, P. Quantifying Bias in a Face Verification System. *CSFM* **2022**, *3*, 6. https://doi.org/10.3390/cmsf2022003006

Academic Editors: Kuan-Chuan Peng and Ziyan Wu

Published: 20 April 2022

1. Introduction

In light of increased reliance on ML in highly consequential applications such as pretrial risk assessment [1,2], occupation classification [3,4], and money lending [5], there is growing concern for the fairness of ML-powered systems [4,6–10]. Unequal performance across individuals and groups subject to a system may have unintended negative consequences for those who experience underperformance [6], potentially depriving them of opportunities, resources, or even freedoms.

Face verification (FV) and face recognition (FR) technologies are widely deployed in systems such as biometric authentication [11], face identification [12], and surveillance [13]. In FV, the input data are two face images and the classifications may be genuine (positive class) or imposters (negative class) [8]. FV/FR typically use a similarity measure (often cosine similarity) applied to a pair of face embeddings produced by the model [12]. There has been recent interest in assessing bias via these face embeddings [10].

Figure 1 presents a low-dimensional depiction of face embeddings generated by FaceNet [12], which clearly groups same-race and same-gender faces closely together, indicating that the model learned to identify the similarities between same-race, same-gender faces. Exploring the connection between embedded clusters of protected groups and biased performance [10] is an open area of research.

In this paper, we (1) identify and quantify sources of bias in a pretrained FaceNet model using statistical and cluster-based measures, and (2) analyze the connection between cluster quality and biased performance.

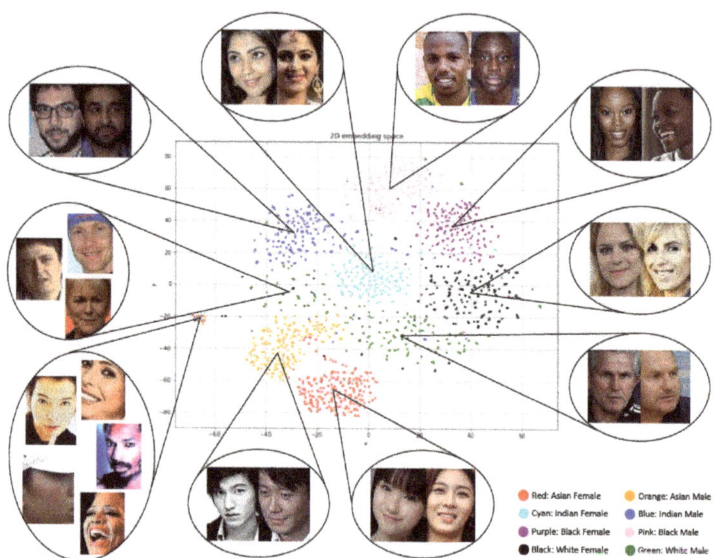

Figure 1. A two-dimensional t-SNE [14] visualization of Balanced Faces in the Wild (BFW) [8] embeddings, colored by race and gender. Clusters roughly correspond to race and gender, with varied densities (e.g., Asian clusters are tighter than White clusters). Note that t-SNE embeddings are not completely representative of actual relationships due to information loss during dimensionality reduction.

2. Related Work

2.1. Sources of Bias

We define bias in an ML system as follows. For a more complete discussion of sources of bias, see the work by Suresh and Guttag [15].

Historical Bias arises when injustice in the world conflicts with values we want encoded in a model. Since systemic injustice creates patterns reflected in data, historical bias can exist despite perfect sampling and representation.

Representation Bias arises when training data under-represent a subset of the target population and the model fails to optimize for the under-represented group(s).

Measurement Bias arises when data are a noisy proxy for the information we desire, e.g., in FV, camera quality and discretized race categories contribute to measurement bias.

Aggregation Bias arises when inappropriately using a "one-size-fits-all" model on distinct populations, as a single model may not generalize well to all subgroups.

Evaluation Bias arises when the evaluation dataset is not representative of the target population. An evaluation may purport good performance, but miss a disparity for populations under-represented in the benchmark dataset.

Deployment Bias arises from inconsistency between the problem that a model is intended to solve and how it is used to make decisions in practice, as there is no guarantee that measured performance and fairness will persist.

2.2. Statistical Fairness Definitions

We first identify attributes of the data for which the system must perform fairly. An attribute may be any qualitative or quantitative descriptor of the data, such as name, gender, or image quality for a face image. A "sensitive" attribute defines a mapping to advantaged and disadvantaged groups [6], breaking a dataset into "unprotected" and "protected" groups. For example, if race is the sensitive attribute, the dataset is broken into an unprotected group, White faces, and protected groups, other-race faces.

We define fairness according to the equal metrics criteria [6,15–18]: a fair model yields similar performance metric results for protected and unprotected subgroups. Other fairness definitions include group-independent predictions [6,15,19,20] (a fair model's decision is not influenced by group membership with respect to a sensitive attribute), individual fairness [6,15,21–23] (individuals who are similar with respect to their attributes have similar outcomes), and causal fairness [6,15,24–26] (developing requirements on a causal graph that links data/attributes to outcomes).

We quantify fairness according to the equal metrics definition using statistical fairness metrics (see Table 1). The metrics use the definitions represented by the confusion matrix in Table 3 of Verma and Rubin [7].

Table 1. Selected statistical fairness metrics. Notation [7,16]: **A**—sensitive attribute, **Y**—actual classification, **d**—predicted classification, and **S**—similarity score. * PPV/NPV: Positive (Negative) Predictive Value.

Metric	Description	Definition	References
Overall Accuracy Equality	Equal prediction accuracy across protected and unprotected groups	$P(d = Y\|A_1) =$ $P(d = Y\|A_2) = \cdots =$ $P(d = Y\|A_N)$	Berk et al. [27] Mitchell et al. [6] Verma and Rubin [7]
Predictive Equality	Equal FPR across protected and unprotected groups	$P(d = 1\|Y = 0, A_1) =$ $P(d = 1\|Y = 0, A_2) = \cdots =$ $P(d = 1\|Y = 0, A_N)$	Chouldechova [17] Corbett-Davies et al. [18] Mitchell et al. [6] Verma and Rubin [7]
Equal Opportunity	Equal FNR across protected and unprotected groups	$P(d = 0\|Y = 1, A_1) =$ $P(d = 0\|Y = 1, A_2) = \cdots =$ $P(d = 0\|Y = 1, A_N)$	Chouldechova [17] Hardt et al. [16] Kusner et al. [24] Mitchell et al. [6] Verma and Rubin [7]
Conditional Use Accuracy Equality	Equal PPV and NPV * across protected and unprotected groups	$P(Y = 1\|d = 1, A_1) =$ $P(Y = 1\|d = 1, A_2) = \cdots =$ $P(Y = 1\|d = 1, A_N)$ AND $P(Y = 0\|d = 0, A_1) =$ $P(Y = 0\|d = 0, A_2) = \cdots =$ $P(Y = 0\|d = 0, A_N)$	Berk et al. [27] Mitchell et al. [6] Verma and Rubin [7]
Balance for the Positive Class	Equal avg. score S for the positive class across protected and unprotected groups	$AVG(Y = 1\|A_1) =$ $AVG(Y = 1\|A_2) = \cdots =$ $AVG(Y = 1\|A_N)$	Kleinberg et al. [28] Mitchell et al. [6] Verma and Rubin [7]
Balance for the Negative Class	Equal avg. score S for the negative class across protected and unprotected groups	$AVG(Y = 0\|A_1) =$ $AVG(Y = 0\|A_2) = \cdots =$ $AVG(Y = 0\|A_N)$	Kleinberg et al. [28] Mitchell et al. [6] Verma and Rubin [7]

2.3. Bias in the Embedding Space

Instead of solely considering model performance across protected and unprotected groups, Gluge et al. [10] assess bias in FV models by investigating the face embeddings produced by the model. The intuition behind this approach is that the "other-race effect" observed in human FV, where people are able to distinguish between same-race faces better than other-race faces, may have an analog in machine FV that is observable in how a model clusters face embeddings according to sensitive attributes such as race, gender, or age.

Gluge et al. [10] attempt to measure bias with respect to a sensitive attribute by quantifying how well embeddings are clustered according to that attribute. They hypothesize that a "good" clustering of embeddings (i.e., well-separated clusters) into race, gender, or age groups may indicate that the model is very aware of race, gender, or age differences, allowing for discrimination based on the respective attribute. They investigate the connection between quality of clustering and bias using cluster validation measures.

Their results do not support a connection between well-defined sensitive attribute clusters and bias; rather, they suggest that a worse clustering of embeddings into sensitive attribute groups yields biased performance (i.e., unequal recognition rates across groups).

They conjecture that between-cluster separation (i.e., how well race, gender, or age groups are separated from each other) may be less important than the within-cluster distribution of embeddings (i.e., how well each individual race, gender, or age group is clustered), intuiting that a cluster's density indicates how similar or dissimilar its embeddings are according to their separation from each other. Thus, a dense cluster may purport false matches more frequently than a less dense cluster. We extend [10] by investigating this conjecture.

3. Method

We experiment using an FV pipeline to evaluate FaceNet [12] on four benchmark datasets. We quantify bias according to the "equal metrics" fairness definition with several distinct statistical fairness metrics, revealing representation bias. We then evaluate clustered embeddings with respect to race and gender groups using clustering metrics and visualizations, revealing aggregation bias. Using the statistical and cluster-based analyses, we draw conclusions on the connection between the clustering of faces into protected and unprotected groups and disparity in model performance between these groups. Figure 2 provides an overview of our method.

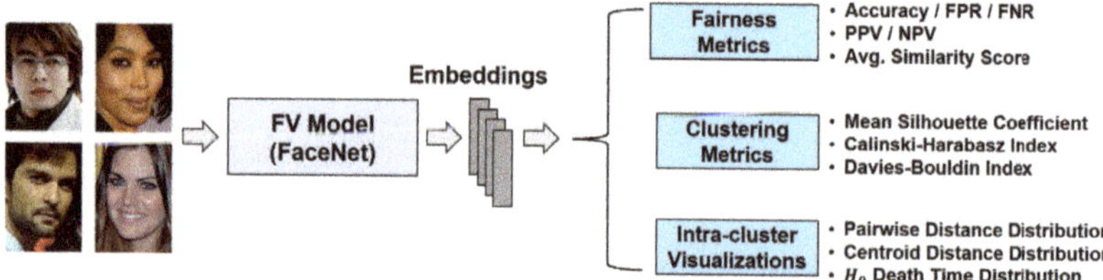

Figure 2. An overview of our approach. We use diverse face datasets to assess bias in FaceNet [12] by leveraging the face embeddings that it produces for various fairness experiments.

3.1. FV Pipeline

We use MTCNN [29] for face detection and a facenet-pytorch Inception V1 model (https://github.com/timesler/facenet-pytorch, accessed on 28 February 2022), cutting out the final, fully connected layer from the FaceNet model so that it produces face embeddings. The constructed pipeline follows.

1. Pass a pair of face images to MTCNN to crop them to bounding boxes around the faces (we discard data where MTCNN detects no faces). Each input pair has an "actual classification" of 1 (genuine) or 0 (imposter).
2. Pass each cropped image tensor into the model (FaceNet, for our experiments) to produce two face embeddings.
3. Compute the cosine similarity between the two embeddings (the "similarity score").
4. Use a pre-determined threshold (the threshold is determined according to a false accept rate (FAR) of 0.05 on a 20% heldout validation set; all datasets have no overlap between people in the testing and validation splits) to produce a "predicted classification" of 1 (genuine) or 0 (imposter).

As detailed in [12], FaceNet is trained using triplet loss on the VGGFace2 dataset [30], comprising faces that are 74.2% White, 15.8% Black, 6.0% Asian, and 4.0% Indian, with 59.3% male and 40.7% female [30].

3.2. Datasets

We run experiments on four benchmark datasets: Balanced Faces in the Wild (BFW) [8], Racial Faces in the Wild (RFW) [31–34], Janus-C [35], and the VGGFace2 [30] test set. Details for each dataset are provided in Table 2.

Table 2. The four benchmark datasets that we use in our experiments. Faces/ID is the average number of faces per ID. * VGG Test represents the VGGFace2 test set.

Dataset	# IDs	Faces/ ID	Attributes	Notes
BFW	800	25	Race, Gender	Equal balance for race and gender
RFW	12,000	6.7	Race	Equal balance for race
IJBC	3531	6	Skin Tone, Gender	Occlusion, occupation diversity
VGG Test *	500	375	Gender	Variation in pose and age

We discuss results primarily for BFW experiments because the dataset is balanced for race and gender. Balance in the sensitive attributes allows valid comparison between results for protected and unprotected groups. BFW comes with pre-generated face pairs with a ratio of 47:53 positive to negative pairs. However, we generate our own positive and negative pairs in order to control holding out 20% of people in the dataset for a validation set.

Table 3 shows the breakdown of our positive and negative pairs by race/gender subgroups for the BFW testing split. Ratios for the validation set are similar. Positive and negative pairs have same-race and same-gender faces. The supplemental material documents pair generation for RFW, Janus-C, and VGGFace2.

We use race and gender as sensitive attributes to examine race, gender, and intersectional race/gender biases [9] in our FV system. The race attribute encompasses four groups (Asian, Indian, Black, and White) consistent across all datasets with a "race" attribute.

Table 3. The percentage of positive and negative pairs per subgroup for the BFW testing split. Ratios for the validation set are similar.

Female	Asian	Indian	Black	White
% positive	25	25	25	25
% negative	75	75	75	75
Male	**Asian**	**Indian**	**Black**	**White**
% positive	25	25	25	25
% negative	75	75	75	75

3.3. Statistical Fairness

To quantify bias according to the "equal metrics" fairness definition, we use nine statistical fairness metrics to evaluate FaceNet model performance on protected and unprotected groups for each sensitive attribute across the four benchmark datasets. We generate bootstrap confidence intervals for all metric results [36].

We compare results between the protected and unprotected groups of each sensitive attribute to identify inequality in model performance, and present seven of the statistical fairness metric results on BFW in this paper (see Table 1 for details). The supplemental material documents results for additional metrics and datasets.

3.4. Cluster-Based Fairness

We extend Gluge et al. [10] by evaluating clustered embeddings to illuminate any connection between sensitive-attribute cluster quality and model performance for protected and unprotected subgroups. For example, we may consider face embeddings from the

BFW dataset to be clustered according to race (four clusters), gender (two clusters), or race/gender (eight clusters). Figure 1 provides a low-dimensional depiction of the BFW embedding space, where groups are distinguished by race/gender.

Based on the findings of [10], we hypothesize a connection between the quality of embedded clusters and model performance, where dense clustering for a particular subgroup is linked to poor performance on that group. Intuition suggests that dense clustering indicates high model confidence in the group affiliation of embeddings within that cluster, but lesser ability to distinguish between individuals within the cluster compared to a less dense group of embeddings. We evaluate clustered embeddings through (1) clustering metrics, and (2) intra-cluster visualizations.

Clustering Metrics We employ the following three metrics [10] to assess embedding space partitioning into clusters according to each sensitive attribute.

- Mean silhouette coefficient [37]: A value in the range $[-1, 1]$ indicating how similar elements are to their own cluster. A higher value indicates that elements are more similar to their own cluster and less similar to other clusters (good clustering).
- Calinski–Harabasz index [38]: The ratio of between-cluster variance and within-cluster variance. A larger index means greater separation between clusters and less within clusters (good clustering).
- Davies–Bouldin index [39]: A value greater than or equal to zero aggregating the average similarity measure of each cluster with its most similar cluster, judging cluster separation according to their dissimilarity (a lower index means better clustering).

Intra-Cluster Visualizations To observe whether or not there is inequality in the embedded cluster quality of protected and unprotected groups, we produce intra-cluster visualizations and compare clusters using pairwise distance distribution, centroid distance distribution, and persistent homology H_0 death time distribution [40,41].

4. Experiments

4.1. Statistical Fairness Metrics

Figure 3 presents statistical fairness metric results for BFW race and gender subgroups; the supplemental materials include complete results for all datasets.

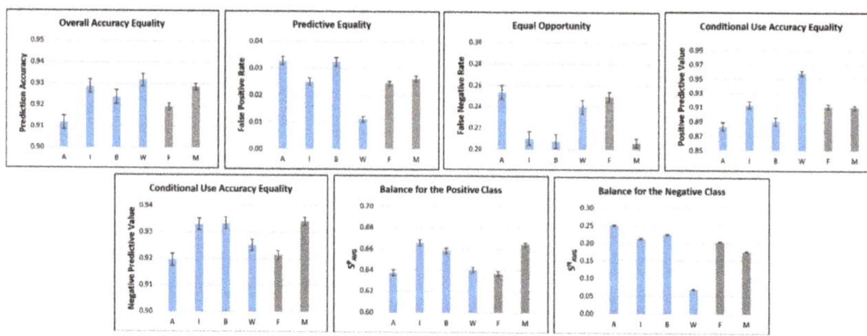

Figure 3. Statistical fairness metric results for BFW race and gender subgroups. See Table 1 for metric descriptions. Blue bars denote race subgroups; gray bars denote gender subgroups. A = Asian; I = Indian; B = Black; W = White; F = Female; M = Male.

While results do not consistently favor one race group, a pattern of bias emerges when considering each metric's implications. Prediction accuracy for Asian faces is lower, but no single race group exhibits significantly better performance than the rest (there is overlap between the confidence intervals of Indian, Black, and White faces). The same observation applies to FNR (lower FNR is better; the Indian and Black confidence intervals overlap) and NPV (higher NPV is better; the Indian and Black confidence intervals overlap). However, FPR and PPV tell a different story.

The model has a low FPR for White faces compared to other race groups, indicating more confidence in White non-matches than for other-race faces. A similar observation is made for PPV; the model is considerably more precise in determining genuine White face pairs compared to other races. The statistics on average similarity score for the positive and negative classes provide an explanation for these results.

Average similarity scores for genuine pairs across race groups are relatively similar ($\sim$0.03 range), but not for imposter pairs ($\sim$0.18 range). Low average similarity scores for White imposter pairs indicate that the model separates non-match White faces very well, hence its confidence in identifying imposter White face pairs (low FPR). Some metrics do not reveal this bias due to comparable average similarity scores across races for genuine pairs; the model is approximately equally confident in identifying genuine pairs for all races, as supported by a similar FNR across race groups.

The inequality in average similarity scores for imposter pairs means that the model learned to distinguish White faces much better than other-race faces, possibly due to encountering significantly more White faces than other-race faces during training. Thus, we identify representation bias as the first form of bias affecting FaceNet. The consistently poor performance on Asian faces, less represented in the training data, supports representation bias. However, despite having the least representation, the metrics indicate better model performance on Indian as compared to Asian faces, hinting that additional biases may be present.

Results for gender subgroups show a performance gap favoring the unprotected (male) vs. protected (female) gender group. However, the performance gaps for female vs. male faces are not as drastic as those for White vs. other-race faces (e.g., balance for the negative class). The lower average similarity score for imposter male faces and higher average similarity score for genuine male faces supports the model's higher confidence in identifying genuine male face pairs (lower FNR). Differences in FPR are insignificant (confidence intervals overlap). The bias in average similarity scores appears in a higher prediction accuracy for male as compared to female face pairs.

We conclude that the gender results are a less extreme example of representation bias, supported by the race and gender breakdown of the training dataset, which is more skewed for race than for gender subgroups.

4.2. Clustering Metrics

We assess embedding clusters using (1) the clustering metrics described in Section 3.4, calculated for each sensitive attribute, and (2) intra-cluster visualizations. Table 4 shows results for BFW; results for other datasets are available in the supplemental material.

Table 4. Clustering metric results for BFW. ↑ means that a higher value indicates better clustering and ↓ means that a lower value indicates better clustering.

Metric	Gender	Race	Both
MS↑	0.034	0.091	0.103
CH↑	280	572	444
DB↓	7.55	4.36	3.98

The trend in mean silhouette coefficient, which quantifies the similarity of elements to their own cluster, appears to vary with the number of clusters per sensitive attribute (i.e., attributes with more clusters have a higher mean silhouette coefficient). Results for the Davies–Bouldin index follow the same pattern, indicating that race/gender clusters are best separated according to similarity, followed by race clusters and then gender clusters.

Results for the Calinski–Harabasz index, quantifying the ratio of between-cluster variance and within-cluster variance, differ. A higher index for race compared to race/gender means that mixed-gender race clusters are better separated than single-gender race/gender clusters. This result indicates that gender clusters within a race are close together compared to the distance between racial groups, a property that is visualized in Figure 1.

While these metrics provide a thorough summary of embeddings clustered by sensitive attributes, they do not help us to understand how protected and unprotected groups within each sensitive attribute are clustered.

4.3. *Intra-Cluster Fairness Visualizations*

We use intra-cluster visualizations to observe within-group clustering inequality between protected and unprotected groups in order to identify a potential connection between cluster quality and statistical metric performance.

For each intra-cluster distribution visualization, we perform two-sided independent two-sample t-tests for every combination of two subgroups in order to identify whether or not the means of two subgroups' distributions are significantly different. (Our null hypothesis for every t-test is that there is no difference in sample mean between the distributions for two subgroups. We accept an alpha level of 0.05 to determine statistical significance.) We perform Dunn–Šidák correction (for BFW, we account for twenty-one null hypotheses comprising all two-subgroup combinations of race and gender subgroups) of the p-values for each dataset to counteract the multiple comparison problem. Corrected p-values of the t-tests for BFW subgroup pairs are documented in the supplemental material.

4.3.1. Pairwise Distance Distribution

Figure 4 depicts a probability density distribution for within-subgroup pairwise distances for BFW race and gender subgroups.

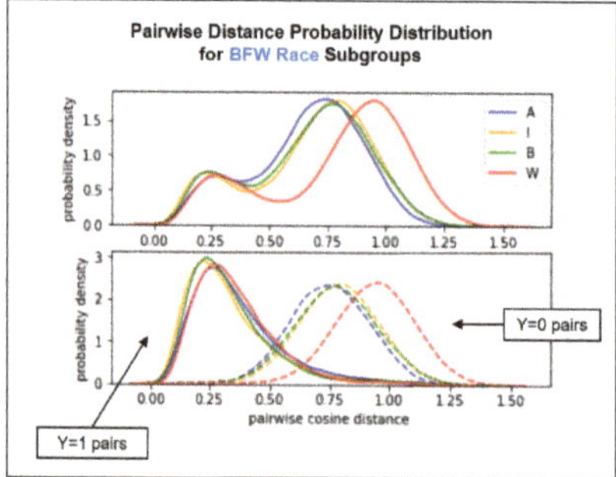

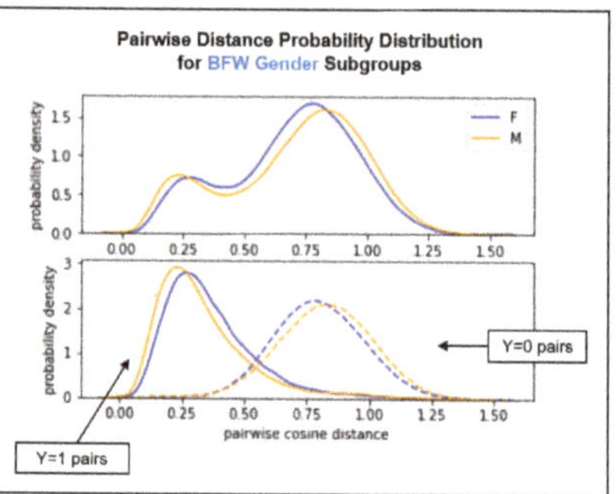

Figure 4. Pairwise distance distribution for BFW race (**left**) and gender (**right**) subgroups. Top plots include all pairs for each subgroup and bottom plots include distinct curves for genuine pairs (solid) and imposter pairs (dashed) for each subgroup.

The White subgroup's negative class plot has a distinct rightward shift compared to other subgroups ($p < 0.05$ for $W \times A$, $W \times I$, and $W \times B$ t-tests), supporting the lower average similarity score for imposter White pairs seen in Figure 3. Consequently, the optimal classification threshold varies by race group; the overlap between the positive and negative class curves for White faces is further right than for other races. Thus, the **average** threshold will be lower than optimal for Asian, Indian, and Black face pairs, leading to more frequent false positives (supported by Figure 3).

We conclude that aggregation bias is present because the classifier relies on one aggregated, sub-optimal threshold for all subgroups [8]. Although the difference between the pairwise distance distributions of gender subgroups is smaller, it is not supported by an insignificant p-value ($p < 0.05$).

4.3.2. Centroid Distance Distribution

Figure 5 depicts a probability density distribution of embedding distances from the centroids of their respective race and gender subgroups for BFW. We use this as a supplementary visualization for within-cluster distances.

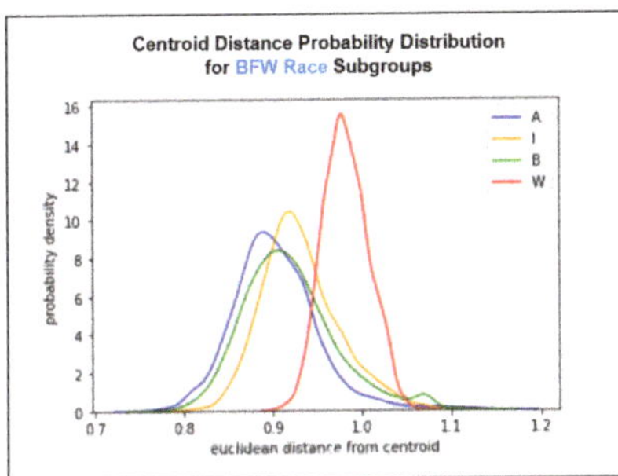

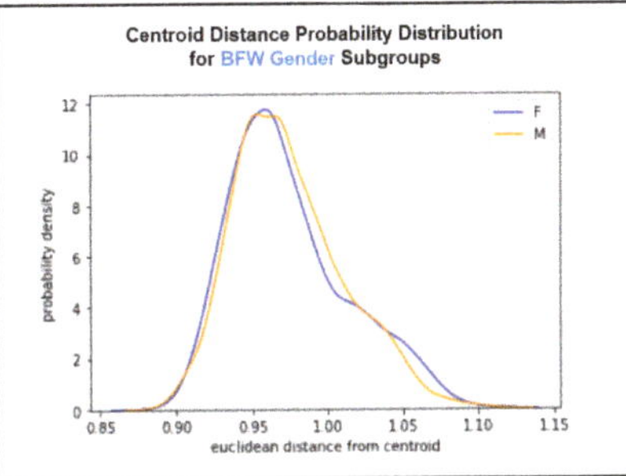

Figure 5. Centroid distance distribution for BFW race subgroups (**left**) and BFW gender subgroups (**right**).

The centroid distance distributions for race subgroups tell a story similar to the pairwise distance distributions, but slightly more nuanced. Faces are uniformly distributed significantly further from the White centroid than in other race groups ($p < 0.05$ for $W \times A$, $W \times I$, and $W \times B$ t-tests). The behavior of Euclidean distance in high-dimensional space [42] suggests that the rightward shift of the White subgroup's plot indicates that White face embeddings are distributed less densely than other race groups. The plots for gender subgroups indicate comparable cluster densities ($p > 0.05$). Thus, centroid distance distribution supports findings from pairwise distance distribution by confirming that White embeddings are better separated than other-race embeddings. It also supports the findings from statistical metrics by demonstrating that there is less inequality between gender clusterings as compared to race clusterings.

4.3.3. Persistent Homology

Our final experiment conducts a more rigorous analysis of the high-dimensional geometry of embedding clusters using persistent homology [40,41], which investigates qualitative information about the structure of data and is suited to high-dimensional, noisy data. Figure 6 depicts density plots for death times of the 0th homology class (H_0) [43] for BFW race and gender subgroups in order to observe trends in the evolution of connected components. "Death time" indicates how many timesteps pass before a connected component "dies" (becomes connected with another component). Thus, death times of connected components is an indicator of the distance between embeddings in the embedding space (i.e., earlier death times indicate that embeddings are generally closer together).

H_0 death times for White face embeddings tend to be later than other race groups ($p < 0.05$ for $W \times A$, $W \times I$, and $W \times B$ t-tests), indicating that White embeddings are more dispersed in the embedding space. The other race groups have peak death times that are taller and earlier than the White race group. The shorter and wider peak for the White subgroup means that there is more variety (higher variance) in H_0 death times, rather than the consistent peak around 0.8 with less variance for other race groups. This shows that there is more variance for White face distribution in the embedding space compared to other race groups, a trend that was not present in the centroid distance distribution for race groups, which showed four bell-shaped density plots. Thus, our analysis of the (H_0) death times supports previous findings that the White race group is clustered differently to other race groups. We note that there is less inequality in H_0 death times for female vs. male faces, despite our p-value indicating that this discrepancy may be significant ($p < 0.05$).

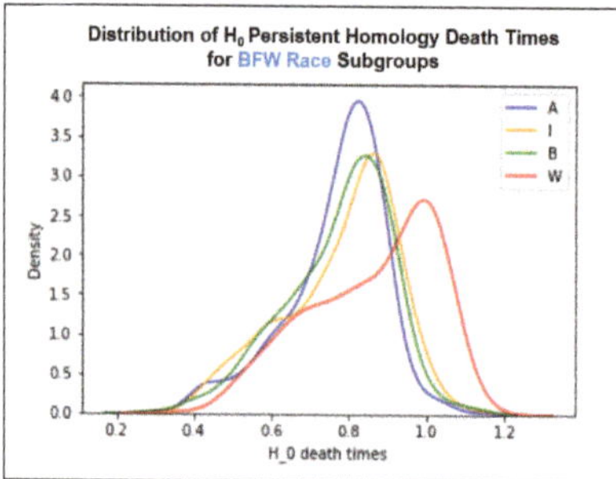

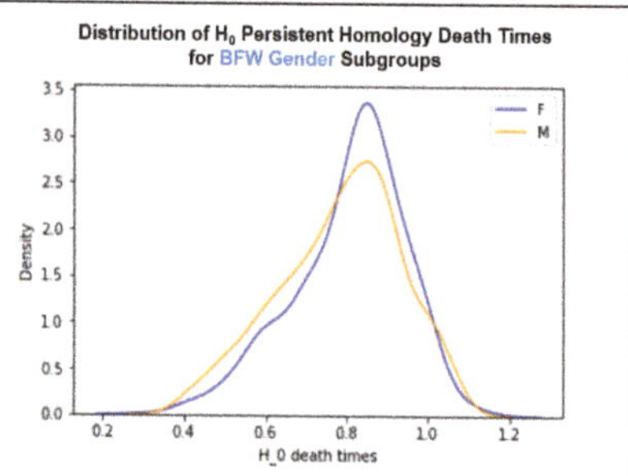

Figure 6. Distribution of persistent homology class 0 (H_0) death times for BFW race (**left**) and gender (**right**) subgroups.

5. Conclusions

We quantify bias in a FaceNet FV system with statistical fairness metrics and clustered embedding evaluations. Unequal statistical metric performance for protected and unprotected race groups reflects representation inequality in the training data, implicating representation bias. However, superior prediction accuracy for some less-represented race groups (e.g., better performance on Indian faces than Asian faces) demonstrates that representation bias is not the only bias present.

Pairwise distance distributions and unequal "balance for the positive/negative class" statistical metrics indicate that the optimal classification threshold varies by race group. Thus, the aggregated classification threshold is skewed lower than optimal for protected race groups, identifying the presence of aggregation bias in the FaceNet FV system.

We demonstrate correspondence between poorly clustered subgroups and those with the best statistical metric performance, supporting our hypothesis that worse clustering may result in less bias. We thus support the intuition that the model learns to distinguish between faces in less dense clusters better than between faces in more dense clusters.

In summary, the model was optimized to perform best on White and male faces due to representation and aggregation bias, resulting in a less dense clustering of unprotected groups in the embedding space. We conclude that FaceNet underperforms on protected demographic groups because, as denser clustering shows, it is less sensitive to differences between facial characteristics within those groups.

Our experiments implicate cluster quality as an apparent indicator of bias, but do not prove causality. We identify causal fairness as an area of future investigation to supplement this work [25]. We also believe that conducting a more rigorous clustering analysis using persistent homology (i.e., quantifying the difference between persistence diagrams) would strengthen the results presented here. Finally, we see potential in applying the metrics used in this paper to multi-class classification problems (namely, FR instead of FV) in both open- and closed-world settings.

The Appendixes A–D provides results from experiments not detailed in the main paper. We first document positive and negative pair generation for Racial Faces in the Wild (RFW) [31], Janus-C [35], and the VGGFace2 [30] test set. We then include results from statistical fairness metrics, clustering metrics, and intra-cluster visualization for Balanced Faces in the Wild (BFW) [8], RFW, Janus-C, and the VGGFace2 test set.

Author Contributions: Conceptualization, M.F., P.K., J.M., K.K. and P.T.-C.; methodology, M.F. and J.M.; software, M.F., P.K. and J.M.; validation, M.F., P.K., J.M. and K.K.; formal analysis, M.F.; investigation, M.F.; resources, P.K., J.M. and K.K.; data curation, M.F.; writing—original draft preparation, M.F.; writing—review and editing, M.F., P.K., J.M. and K.K.; visualization, M.F.; supervision, P.K., J.M. and K.K.; project administration, P.K.; funding acquisition, P.T.-C. All authors have read and agreed to the published version of the manuscript.

Funding: DISTRIBUTION STATEMENT A. Approved for public release. Distribution is unlimited. This material is based upon work supported by the Department of Defense under Air Force Contract No. FA8702-15-D-0001. Any opinions, findings, conclusions, or recommendations expressed in this material are those of the author(s) and do not necessarily reflect the views of the Department of Defense.

Institutional Review Board Statement: Not applicable.

Informed Consent Statement: Not applicable.

Data Availability Statement: The BFW dataset contains third-party data and is available upon registration/request from the original dataset authors at https://forms.gle/3HDBikmz36i9DnFf7 (accessed on 28 February 2022). The RFIW dataset contains third-party data and is available upon request from the original dataset authors at http://whdeng.cn/RFW/testing.html (accessed on 28 February 2022). The VGGFace2 test set contains third-party data and must be requested from the original dataset authors (https://doi.org/10.1109/FG.2018.00020, accessed on 28 February 2022). The Janus-C dataset contains third-party data and is available upon request from NIST (not corresponding author) at https://www.nist.gov/itl/iad/ig/ijb-c-dataset-request-form (accessed on 28 February 2022).

Acknowledgments: The authors would like to thank Joseph Robinson and Mei Wang for granting access to the BFW and RFW datasets, respectively. This product contains or makes use of the following data made available by the Intelligence Advanced Research Projects Activity (IARPA): IARPA Janus Benchmark C (IJB-C) data detailed at Face Challenges homepage (https://www.nist.gov/programs-projects/face-challenges, accessed on 28 February 2022).

Conflicts of Interest: The authors declare no conflict of interest.

Abbreviations

The following abbreviations are used in this manuscript:

BFW	Balanced Faces in the Wild
CH	Calinski–Harabasz Index
DB	Davies–Bouldin Index
FNR	False Negative Rate
FPR	False Positive Rate
FR	Face Recognition
FV	Face Verification
IJBC	IARPA Janus Benchmark C
ML	Machine Learning
MS	Mean Silhouette Coefficient
MTCNN	Multi-Task Cascaded Convolutional Networks
NPV	Negative Predictive Value
PPV	Positive Predictive Value
RFW	Racial Faces in the Wild
t-SNE	t-distributed Stochastic Neighbor Embedding

Appendix A. Pair Generation

Racial Faces in the Wild Table A1 displays the breakdown of positive and negative pairs for the RFW testing split for each race subgroup. Positive and negative pairs are same-race faces (there is no gender attribute for this dataset).

Table A1. The test set percentages of positive and negative pairs generated per subgroup for RFW.

	Asian	Indian	Black	White
% positive	25.0	25.0	25.2	25.0
% negative	75.0	75.0	74.8	75.0

Janus-C Table A2 details the Janus-C test set's positive and negative pairs across skin tone and gender subgroups. All pairs are same-skin-tone and same-gender faces. Because Janus-C is not balanced over sensitive attributes, we had to vary positive and negative pair generation for each skin tone and gender subgroup. The drastically different number of faces across skin tones and genders make it difficult to achieve parity in the number of pairs for these subgroups while maintaining a large enough sample for testing. This should be considered when interpreting Janus-C results.

Table A2. The test set percentages of positive and negative pairs generated per subgroup for Janus-C.

Female	**1**	**2**	**3**	**4**	**5**	**6**
% positive	54.9	40.6	36.5	14.9	14.4	7.1
% negative	45.1	59.4	63.5	85.1	85.6	92.9
Male	**1**	**2**	**3**	**4**	**5**	**6**
% positive	54.7	37.3	29.5	13.7	8.6	5.5
% negative	45.3	62.7	70.5	86.3	91.4	94.5

VGGFace2 Test Set Table A3 shows the breakdown across gender subgroups of positive and negative pairs for the VGG testing split. All pairs are same-gender faces (VGGFace2 does not have a race attribute). The VGGFace2 test set is not balanced over its sensitive attribute, so we had to vary positive and negative pair generation by gender subgroup. Because VGGFace2 has less inequality than Janus-C in number of faces per subgroup, we achieved positive to negative pair ratios much closer to 25:75.

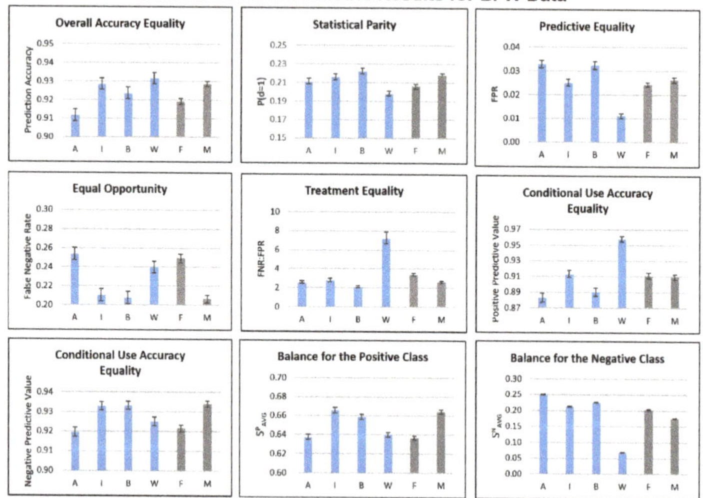

Figure A1. Statistical fairness metric results for BFW subgroups. A = Asian; I = Indian; B = Black; W = White; F = Female; M = Male.

Table A3. The test set percentages of positive and negative pairs generated per subgroup for the VGGFace2 test set.

	Female	**Male**
% positive	23.6	29.6
% negative	76.4	70.4

Appendix B. Statistical Fairness Metric Experiments

Figure A1 documents statistical metric results for BFW data that are not included in the main paper, while Figures A2 and A3 document results for RFW and VGGFace2, respectively.

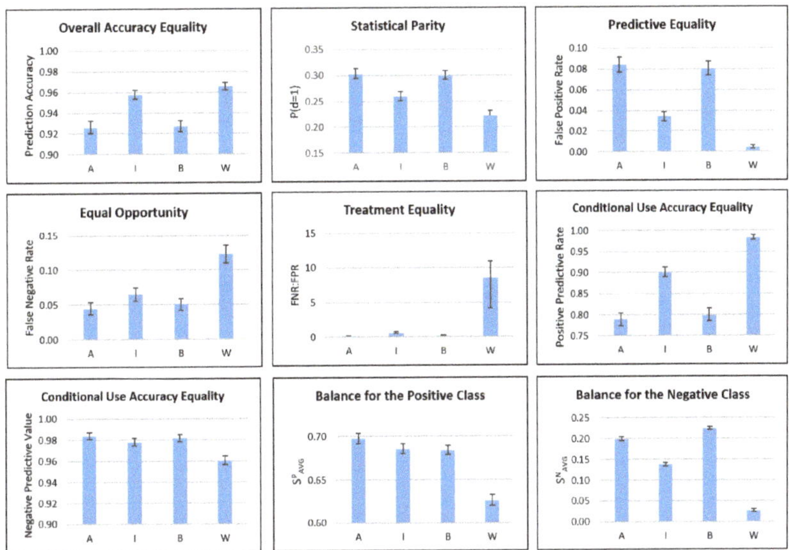

Figure A2. Statistical fairness metric results for RFW race subgroups. A = Asian; I = Indian; B = Black; W = White.

Statistical Fairness Metric Results for the VGGFace2 Test Set

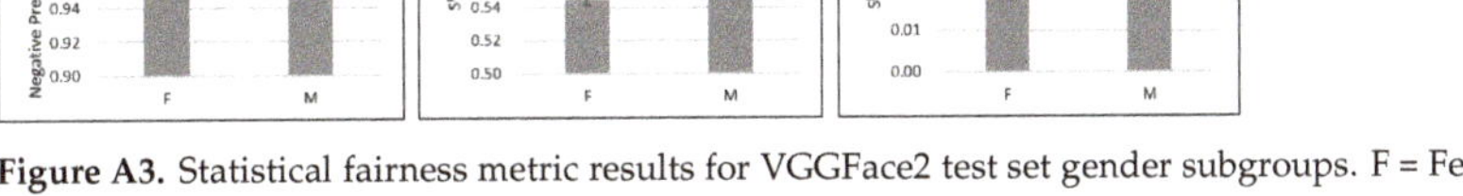

Figure A3. Statistical fairness metric results for VGGFace2 test set gender subgroups. F = Female; M = Male.

We attempt to take advantage of the skin tone attribute in Janus-C to assess performance deficits relating specifically to skin color. We hypothesize that an FV system may perform worse on darker faces than lighter faces due to factors such as lighting or image quality. We attempt to measure this by running two experiments: one with a Gaussian blur filter applied to the images and one without.

We compare blurred and non-blurred image results, expecting a greater drop in performance for blur with darker skin tones, indicating that darker faces likely appear in lower-quality images to begin with (a form of measurement bias). Figure A4 documents the results of these Janus-C experiments. We do not include these results in the main paper because (1) the inconsistent ratios of positive and negative pairs make it difficult to compare results across skin tones, and (2) we do not see significant performance changes after adding blur (the changes fall within the margin of error).

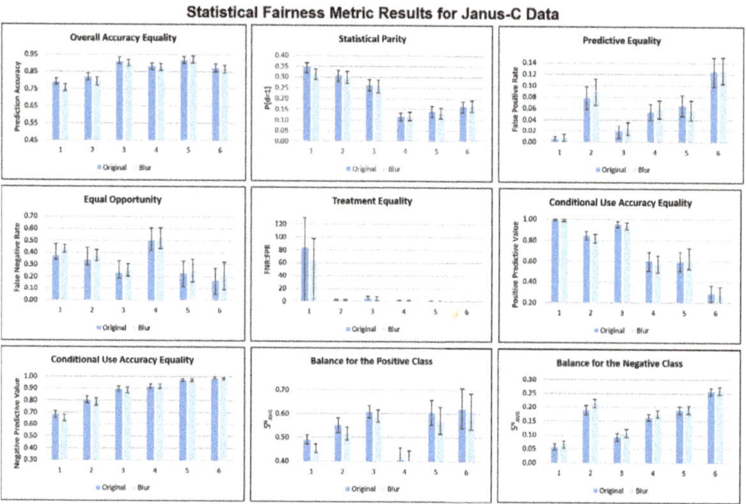

Figure A4. Statistical fairness metric results for Janus-C skin tone subgroups. Dark blue bars represent original data; light blue bars represent blurred data. Skin tone groups are labelled from 1 (lightest skin) to 6 (darkest skin).

Appendix C. Clustering Metrics

Tables A4–A6 display clustering metric results for RFW, VGGFace2, and Janus-C, respectively. As stated in the main paper, these results do not add support to the connection between cluster quality and model performance. However, they provide a quantification of embedding clustering according to various sensitive attributes that is useful for understanding each dataset's clustered embeddings.

Table A4. Clustering metric results for RFW. ↑ means that a higher value indicates better clustering and ↓ means that a lower value indicates better clustering.

Metric	Race
MS↑	0.112
CH↑	1423
DB↓	4.21

Table A5. Clustering metric results for the VGGFace2 test set.

Metric	Gender
MS↑	0.026
CH↑	1835
DB↓	8.44

Table A6. Clustering metric results for Janus-C.

Metric	Gender	Skin Tone
MS↑	0.034	−0.002
CH↑	380	227
DB↓	7.57	7.81

Appendix D. Clustering Visualizations

Figures A5–A7 document intra-cluster visualizations for RFW, VGGFace2, and Janus-C, respectively. For each dataset and sensitive attribute, we include pairwise distance distributions, centroid distance distributions, and persistent homology 0th class death distributions.

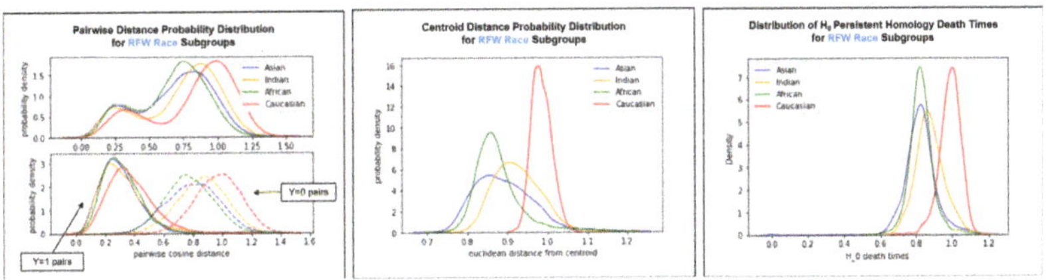

Figure A5. Intra-cluster visualizations for RFW. Pairwise distance distribution (**left**); centroid distance distribution (**middle**); persistent homology 0th class deaths distribution (**right**).

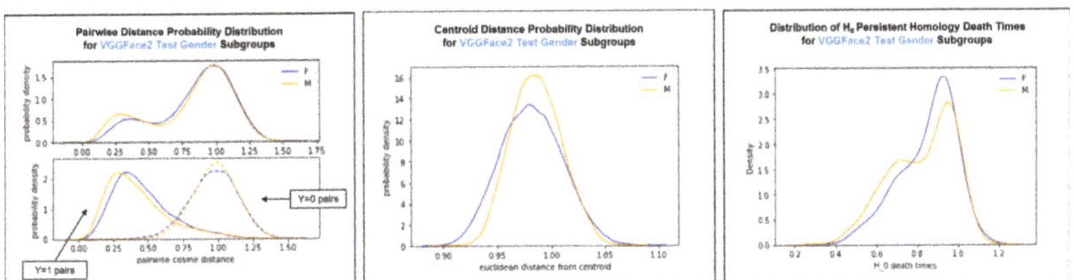

Figure A6. Intra-cluster visualizations for the VGGFace2 test set. Pairwise distance distribution (**left**); centroid distance distribution (**middle**); persistent homology 0th class deaths distribution (**right**).

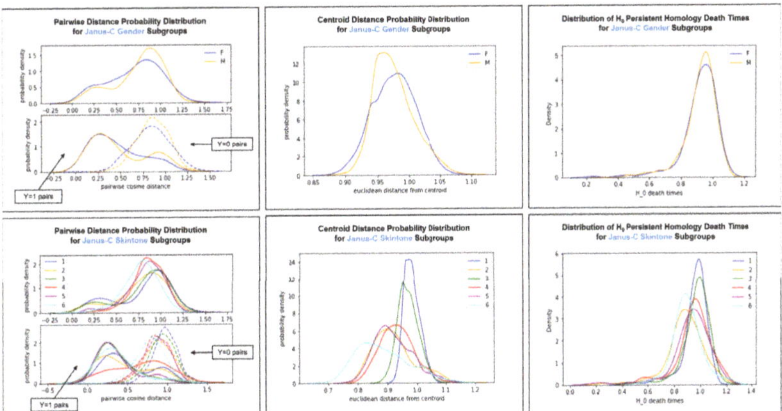

Figure A7. Intra-cluster visualizations for Janus-C. Pairwise distance distribution (**left**); centroid distance distribution (**middle**); persistent homology 0th class deaths distribution (**right**).

Trends in RFW and Janus-C skin tone intra-cluster visualizations are similar to trends in BFW race intra-cluster visualizations; White faces (or lighter faces in Janus-C; skin tone group 1) belong to less dense and more dispersed clusters than other-race faces.

Trends in VGGFace2 and Janus-C gender intra-cluster visualizations are similar to trends in BFW gender intra-cluster visualizations; there is little difference in clustering between male and female faces.

Intra-Cluster Distribution T-Tests

In the main paper, we describe the calculation of p-values for intra-cluster distribution t-tests, used to determine if the means of two subgroups' distributions are significantly different. p-values below the alpha-level of 0.05 validate observations from the intra-cluster visualizations, namely that White faces are less densely clustered in the embedding space than other-race faces. Tables A5–A8 document corrected p-values of the t-tests for BFW, RFW, VGGFace2, and Janus-C subgroup pairs, respectively.

Table A7. Corrected p-values of the 2-sample independent t-test results for BFW race (top) and gender (bottom) subgroup pairs. A: Asian; I: Indian; B: Black; W: White; F: Female; M: Male.

Pairwise Distance Distributions				Centroid Distance Distributions				H_0 Death Time Distributions			
	I	B	W		I	B	W		I	B	W
A	<0.001	<0.001	<0.001	A	<0.001	<0.001	<0.001	A	>0.999	>0.999	<0.001
I	-	<0.001	<0.001	I	-	<0.001	<0.001	I	-	>0.999	<0.001
B	-	-	<0.001	B	-	-	<0.001	B	-	-	<0.001
Pairwise Distance Distributions				**Centroid Distance Distributions**				**H_0 Death Time Distributions**			
	M				M				M		
F	<0.001			F	>0.999			F	>0.03		

Table A8. Corrected p-values of the 2-sample independent t-test results for RFW race subgroup pairs. Top: race subgroup results; bottom: gender subgroup results. A: Asian; I: Indian; B: Black; W: White.

Pairwise Distance Distributions				Centroid Distance Distributions				H_0 Death Time Distributions			
	I	B	W		I	B	W		I	B	W
A	<0.001	<0.001	<0.001	A	<0.001	<0.001	<0.001	A	<0.001	>0.999	<0.001
I	-	<0.001	<0.001	I	-	<0.001	<0.001	I	-	<0.001	<0.001
B	-	-	<0.001	B	-	-	<0.001	B	-	-	<0.001

Table A9. Corrected p-values of the 2-sample independent t-test results for VGGFace2 test set gender subgroup pairs. F: Female; M: Male.

Pairwise Distance Distributions		Centroid Distance Distributions		H_0 Death Time Distributions	
	M		M		M
F	<0.001	F	<0.001	F	0.02

Table A10. Corrected p-values of the 2-sample independent t-test results for Janus-C skin tone (top) and gender (bottom) subgroup pairs. Results are for non-blurred data. Skin tone groups are labelled from 1 (lightest skin) to 6 (darkest skin). F: Female; M: Male.

	Pairwise Distance Distributions						**Centroid Distance Distributions**						**H_0 Death Time Distributions**				
	2	3	4	5	6		2	3	4	5	6		2	3	4	5	6
1	>0.999	>0.999	0.01	0.62	>0.999	1	<0.001	<0.001	<0.001	<0.001	<0.001	1	<0.001	0.81	>0.999	0.13	<0.001
2	-	0.70	<0.001	0.02	>0.999	2	-	<0.001	<0.001	>0.999	<0.001	2	-	<0.001	<0.001	0.22	>0.999
3	-	-	0.30	>0.999	0.54	3	-	-	<0.001	<0.001	<0.001	3	-	-	0.98	0.03	<0.001
4	-	-	-	0.91	<0.001	4	-	-	-	<0.001	<0.001	4	-	-	-	0.98	0.06
5	-	-	-	-	0.003	5	-	-	-	-	<0.001	5	-	-	-	-	0.99

Pairwise Distance Distributions		**Centroid Distance Distributions**		**H_0 Death Time Distributions**	
	M		M		M
F	0.08	F	>0.999	F	>.999

References

1. Monahan, J.; Skeem, J.L. Risk Assessment in Criminal Sentencing. *Annu. Rev. Clin. Psychol.* **2016**, *12*, 489–513. [CrossRef] [PubMed]
2. Christin, A.; Rosenblat, A.; Boyd, D. Courts and Predictive Algorithms. Data & Civil Rights: A New Era of Policing and Justice, 2016. Available online: https://www.law.nyu.edu/sites/default/files/upload_documents/Angele%20Christin.pdf (accessed on 28 February 2022).
3. Romanov, A.; De-Arteaga, M.; Wallach, H.; Chayes, J.; Borgs, C.; Chouldechova, A.; Geyik, S.; Kenthapadi, K.; Rumshisky, A.; Kalai, A.T. What's in a Name? Reducing Bias in Bios without Access to Protected Attributes. *arXiv* **2019**, arXiv:1904.05233.
4. De-Arteaga, M.; Romanov, A.; Wallach, H.; Chayes, J.; Borgs, C.; Chouldechova, A.; Geyik, S.; Kenthapadi, K.; Kalai, A.T. Bias in Bios: A Case Study of Semantic Representation Bias in a High-Stakes Setting. In Proceedings of the Conference on Fairness, Accountability, and Transparency, Atlanta, GA, USA, 29–31 January 2019.
5. Fuster, A.; Goldsmith-Pinkham, P.; Ramadorai, T.; Walther, A. Predictably Unequal? The Effects of Machine Learning on Credit Markets. *SSRN Electron. J.* **2017**, *77*, 5–47. [CrossRef]
6. Mitchell, S.; Potash, E.; Barocas, S.; D'Amour, A.; Lum, K. Prediction-Based Decisions and Fairness: A Catalogue of Choices, Assumptions, and Definitions. *arXiv* **2020**, arXiv:1811.07867.
7. Verma, S.; Rubin, J. Fairness Definitions Explained. In Proceedings of the 2018 IEEE/ACM International Workshop on Software Fairness (FairWare), Gothenburg, Sweden, 29 May 2018; pp. 1–7. [CrossRef]
8. Robinson, J.P.; Livitz, G.; Henon, Y.; Qin, C.; Fu, Y.; Timoner, S. Face Recognition: Too Bias, or Not Too Bias? In Proceedings of the 2020 IEEE/CVF Conference on Computer Vision and Pattern Recognition Workshops (CVPRW), Seattle, WA, USA, 14–19 June 2020; IEEE: Seattle, WA, USA, 2020; pp. 1–10. [CrossRef]
9. Buolamwini, J.; Gebru, T. Gender Shades: Intersectional Accuracy Disparities in Commercial Gender Classification. In Proceedings of the Proceedings of the Conference on Fairness, Accountability, and Transparency, New York, NY, USA, 23–24 February 2018; ACM: New York, NY, USA, 2018.
10. Gluge, S.; Amirian, M.; Flumini, D.; Stadelmann, T. How (Not) to Measure Bias in Face Recognition Networks. In *Artificial Neural Networks in Pattern Recognition*; Schilling, F.P., Stadelmann, T., Eds.; Springer International Publishing: Berlin/Heidelberg, Germany, 2020; pp. 125–137.
11. Bhattacharyya, D.; Ranjan, R. Biometric Authentication: A Review. *Int. J. u- e-Serv. Sci. Technol.* **2009**, *2*, 13–28.
12. Schroff, F.; Kalenichenko, D.; Philbin, J. FaceNet: A Unified Embedding for Face Recognition and Clustering. In Proceedings of the 2015 IEEE Conference on Computer Vision and Pattern Recognition (CVPR), Boston, MA, USA, 7–12 June 2015.
13. Wheeler, F.W.; Weiss, R.L.; Tu, P.H. Face recognition at a distance system for surveillance applications. In Proceedings of the 2010 Fourth IEEE International Conference on Biometrics: Theory, Applications and Systems (BTAS), Washington, DC, USA, 27–29 September 2010; IEEE: Washington, DC, USA, 2010; pp. 1–8. [CrossRef]
14. van der Maaten, L.; Hinton, G. Visualizing Data using t-SNE. *J. Mach. Learn. Res.* **2008**, *9*, 2579–2605.
15. Suresh, H.; Guttag, J.V. A Framework for Understanding Unintended Consequences of Machine Learning. *arXiv* **2020**. arXiv:1901.10002.
16. Hardt, M.; Price, E.; Price, E.; Srebro, N. Equality of Opportunity in Supervised Learning. In Proceedings of the Advances in Neural Information Processing Systems 29, Barcelona, Spain, 5–10 December 2016; Lee, D.D., Sugiyama, M., Luxburg, U.V., Guyon, I., Garnett, R., Eds.; Curran Associates, Inc.: Red Hook, NY, USA, 2016, pp. 3315–3323.
17. Chouldechova, A. Fair Prediction with Disparate Impact: A Study of Bias in Recidivism Prediction Instruments. *Big Data* **2017**, *5*, 153–163. [CrossRef] [PubMed]

18. Corbett-Davies, S.; Pierson, E.; Feller, A.; Goel, S.; Huq, A. Algorithmic Decision Making and the Cost of Fairness. In Proceedings of the 23rd ACM SIGKDD International Conference on Knowledge Discovery and Data Mining, Halifax, NS, Canada 13–17 August 2017; Association for Computing Machinery: Halifax, NS, Canada, 2017; pp. 797–806. [CrossRef]
19. Zemel, R. Learning Fair Representations. In Proceedings of the ICML, Atlanta, GA, USA, 16–21 June 2013; pp. 325–333.
20. Feldman, M.; Friedler, S.A.; Moeller, J.; Scheidegger, C.; Venkatasubramanian, S. Certifying and Removing Disparate Impact. In Proceedings of the 21th ACM SIGKDD International Conference on Knowledge Discovery and Data Mining, Sydney, NSW, Australia, 10–13 August 2015; Association for Computing Machinery: Sydney, NSW, Australia, 2015; pp. 259–268. [CrossRef]
21. Louizos, C.; Swersky, K.; Li, Y.; Welling, M.; Zemel, R. The Variational Fair Autoencoder. *arXiv* **2017**. arXiv:1511.00830.
22. Rothblum, G.N.; Yona, G. Probably Approximately Metric-Fair Learning. In Proceedings of the ICML, Stockholm, Sweden, 10–15 July 2018.
23. Dwork, C.; Hardt, M.; Pitassi, T.; Reingold, O.; Zemel, R. Fairness through awareness. In Proceedings of the 3rd Innovations in Theoretical Computer Science Conference, Cambridge, MA, USA, 8–10 January 2012; Association for Computing Machinery: Cambridge, MA, USA, 2012; pp. 214–226. [CrossRef]
24. Kusner, M.J.; Loftus, J.; Russell, C.; Silva, R. Counterfactual Fairness. In *Advances in Neural Information Processing Systems 30*; Guyon, I., Luxburg, U.V., Bengio, S., Wallach, H., Fergus, R., Vishwanathan, S., Garnett, R., Eds.; Curran Associates, Inc.: Red Hook, NY, USA, 2017; pp. 4066–4076.
25. Kilbertus, N.; Rojas Carulla, M.; Parascandolo, G.; Hardt, M.; Janzing, D.; Schölkopf, B. Avoiding Discrimination through Causal Reasoning. In *Advances in Neural Information Processing Systems 30*; Guyon, I.; Luxburg, U.V., Bengio, S., Wallach, H., Fergus, R., Vishwanathan, S., Garnett, R., Eds.; Curran Associates, Inc.: Red Hook, NY, USA, 2017; pp. 656–666.
26. Nabi, R.; Shpitser, I. Fair Inference On Outcomes. In Proceedings of the Thirty-Second AAAI Conference on Artificial Intelligence, New Orleans, LA, USA, 2–7 February 2018.
27. Berk, R.; Heidari, H.; Jabbari, S.; Kearns, M.; Roth, A. Fairness in Criminal Justice Risk Assessments: The State of the Art. *Sociol. Methods Res.* **2018**, *50*, 3–44. [CrossRef]
28. Kleinberg, J.; Mullainathan, S.; Raghavan, M. Inherent Trade-Offs in the Fair Determination of Risk Scores. *arXiv* **2018**, arXiv:1609.05807.
29. Zhang, K.; Zhang, Z.; Li, Z.; Qiao, Y. Joint Face Detection and Alignment using Multi-task Cascaded Convolutional Networks. *IEEE Signal Process. Lett.* **2016**, *23*, 1499–1503. [CrossRef]
30. Cao, Q.; Shen, L.; Xie, W.; Parkhi, O.M.; Zisserman, A. VGGFace2: A dataset for recognising faces across pose and age. *arXiv* **2018**. arXiv:1710.08092.
31. Wang, M.; Deng, W.; Hu, J.; Tao, X.; Huang, Y. Racial Faces in-the-Wild: Reducing Racial Bias by Information Maximization Adaptation Network. *arXiv* **2019**. arXiv:1812.00194.
32. Wang, M.; Zhang, Y.; Deng, W. Meta Balanced Network for Fair Face Recognition. *IEEE Trans. Pattern Anal. Mach. Intell.* **2021**. [CrossRef]
33. Wang, M.; Deng, W. Mitigate Bias in Face Recognition using Skewness-Aware Reinforcement Learning. *arXiv* **2019**, arXiv:1911.10692.
34. Wang, M.; Deng, W. Deep Face Recognition: A survey. *Neurocomputing* **2021**, *429*, 215–244. [CrossRef]
35. Maze, B.; Adams, J.; Duncan, J.A.; Kalka, N.; Miller, T.; Otto, C.; Jain, A.K.; Niggel, W.T.; Anderson, J.; Cheney, J.; et al. IARPA Janus Benchmark – C: Face Dataset and Protocol. In Proceedings of the 2018 International Conference on Biometrics (ICB), Gold Coast, QLD, Australia, 20–23 February 2018; IEEE: New York, NY, USA, 2018, pp. 158–165. [CrossRef]
36. Orloff, J.; Bloom, J. Bootstrap confidence intervals, 2014. Available online: https://math.mit.edu/~dav/05.dir/class24-prep-a.pdf (accessed on 28 February 2022).
37. Rousseeuw, P.J. Silhouettes: A graphical aid to the interpretation and validation of cluster analysis. *J. Comput. Appl. Math.* **1987**, *20*, 53–65. [CrossRef]
38. Calinski, T.; Harabasz, J. A dendrite method for cluster analysis. *Commun. Stat.-Theory Methods* **1974**, *3*, 1–27. [CrossRef]
39. Davies, D.L.; Bouldin, D.W. A Cluster Separation Measure. *IEEE Trans. Pattern Anal. Mach. Intell.* **1979**, *1*, 224–227. [CrossRef]
40. Chazal, F.; Michel, B. An introduction to Topological Data Analysis: Fundamental and practical aspects for data scientists. *arXiv* **2017**, arXiv:1710.04019.
41. Wasserman, L. Topological Data Analysis. *Annu. Rev. Stat. Appl.* **2018**, *5*, 501–532. [CrossRef]
42. Domingos, P. A few useful things to know about machine learning. *Commun. ACM* **2012**, *55*, 78-87. [CrossRef]
43. Saul, N.; Tralie, C. Scikit-TDA: Topological Data Analysis for Python. 2019. Available online: https://zenodo.org/record/2533369 (accessed on 28 February 2022).

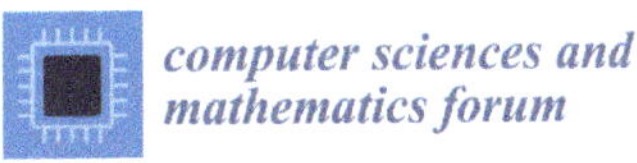
computer sciences and mathematics forum

Proceeding Paper

Super-Resolution for Brain MR Images from a Significantly Small Amount of Training Data †

Kumpei Ikuta [1,*], Hitoshi Iyatomi [1], Kenichi Oishi [2] and on behalf of the Alzheimer's Disease Neuroimaging Initiative ‡

1 Department of Applied Informatics, Graduate School of Science and Engineering, Hosei University, Tokyo 102-8160, Japan; iyatomi@hosei.ac.jp
2 Department of Radiology and Radiological Science, Johns Hopkins University School of Medicin, Baltimore, MD 21205, USA; koishi2@jhmi.edu
* Correspondence: kunpei.ikuta@gmail.com

† Presented at the AAAI Workshop on Artificial Intelligence with Biased or Scarce Data (AIBSD), Online, 28 February 2022.
‡ Data used in preparation of this article were obtained from the Alzheimer's Disease Neuroimaging Initiative (ADNI) database (adni.loni.usc.edu). As such, the investigators within the ADNI contributed to the design and implementation of ADNI and/or provided data but did not participate in analysis or writing of this report. A complete listing of ADNI investigators can be found at https://adni.loni.usc.edu/wp-content/uploads/how_to_apply/ADNI_Acknowledgement_List.pdf.

Abstract: We propose two essential techniques to effectively train generative adversarial network-based super-resolution networks for brain magnetic resonance images, even when only a small number of training samples are available. First, stochastic patch sampling is proposed, which increases training samples by sampling many small patches from the input image. However, sampling patches and combining them causes unpleasant artifacts around patch boundaries. The second proposed method, an artifact-suppressing discriminator, suppresses the artifacts by taking two-channel input containing an original high-resolution image and a generated image. With the introduction of the proposed techniques, the network achieved generation of natural-looking MR images from only ~40 training images, and improved the area-under-curve score on Alzheimer's disease from 76.17% to 81.57%.

Keywords: super-resolution; generative adversarial networks; medical image processing

Citation: Ikuta, K.; Iyatomi, H.; Oishi, K.; on behalf of the Alzheimer's Disease Neuroimaging Initiative Super-Resolution for Brain MR Images from a Significantly Small Amount of Training Data. *CSFM* **2022**, *3*, 7. https://doi.org/10.3390/cmsf2022003007

Academic Editors: Kuan-Chuan Peng and Ziyan Wu

Published: 27 April 2022

1. Introduction

In medical imaging, magnetic resonance imaging (MRI) is commonly used because it can capture the anatomical structure of the human body without exposing subjects to radiation. MRI scanners that generate a strong magnetic field can scan images with a higher spatial and contrast resolution than commonly used scanners. These high-resolution MR images are preferred in both the clinical and research fields since more information can be obtained from a single scanning session, leading doctors to diagnose diseases earlier or computers to analyze images more precisely.

3T MRI scanners provide high spatial resolution and contrast images and are widely used in clinical practice and research studies. Moreover, 7T ultra-high-field scanners are now becoming available for research use, providing ultra-high resolution images that depict fine anatomical structures in unprecedented detail and with higher contrast. Such ultra-high resolution MRI is attractive because it has the potential to capture mild disease-related anatomical changes that are difficult to identify with 3T MRI. Alternatively, obtaining high-definition images with commonly used scanners requires longer scanning times and places a burden on the patient. In such a situation, super-resolving techniques draw attention, which translates low-resolution (LR) MR images to high-resolution (HR) MR images [1].

Resolution enhancing methods for MRI can be categorized into two groups: (1) processing the raw signal from the MRI scanner to improve the resolution to be reconstructed and (2) translating already reconstructed LR images into HR-like images, so-called super-resolution (SR).

From a practical point of view, we chose a post-processing method instead of processing the raw signal from the scanner. The choice is for the following three reasons: (1) MR images are usually stored as rendered image files, while the raw signal data are discarded immediately after each scan. In this approach, an extensive archive of legacy MR images can be used. (2) Post-processing can be used to perform super-resolution. (3) This approach is independent of specific scanner hardware and scan protocols, and can be applied to many MRI contrasts, such as T1-MRI, T2-MRI, diffusion MRI, and functional MRI.

Although these deep-learning (DL) methods, including recent generative adversarial network (GAN) -based ones, have many desirable features from non-DL techniques, they have not yet been able to synthesize images as if they were taken by a high-field scanner. This is because most of their methods are designed to be trained with pairs of ordinary resolution MRI and its shrunken version. Therefore, the conventional SR methods can only learn to translate low-resolution images to normal-resolution images and cannot perform normal-to-high translation, which is an essential demand by clinicians and researchers.

What makes the normal-to-high translation difficult is the limited number of high-definition training images. Deep neural networks, especially GANs, require a large number of training samples to achieve desired performance. Due to the limitation of not having access to images taken by high-end scanners, it is virtually impossible to apply existing DL-based algorithms.

This paper proposes a simple yet effective GAN-based super-resolution method. Compared to the existing DL-based super-resolution methods, the proposed method requires significantly less training of MR images (dozens of data) and generates high-quality SR images. The proposed method comprises two techniques: stochastic patch sampling (SPS) and artifact suppressing discriminator (ASD). The SPS partitions input LR MR images into several smaller patches (i.e., cubes) first. After the partitioning, the ESRGAN-based neural network takes each LR patch as an input, and then outputs the corresponding upscaled HR patch. Here, the ASD eliminates discontinuities in the joints of each patch and generates natural-looking high-resolution images. In our experiments to evaluate the performance of our SR method using 7T MR images of 37 patients, peak signal-to-noise ratio (PSNR) and structural similarity (SSIM) were significantly improved from 16.19 to 26.92 and 0.766 to 0.944, respectively, compared with baseline ESRGAN. In addition, the diagnostic performance of the Alzheimer's disease discriminator trained on super-resolution processed images improved from 80.31% to 83.85%.

2. Related Works

In the last decade, the accuracy of single image super-resolution for general (non-medical) images has increased significantly along with the advancement of DL-based algorithms [2]. Originating from the super-resolution convolutional neural networks (SRCNNs) [3], a very first successful attempt to utilize convolutional neural networks to perform super-resolution, many studies have proposed DL-based SR techniques. Very-deep super-resolution (VDSR) [4] extended SRCNN with a deeper network to improve the accuracy. Enhanced deep residual networks for single image super-resolution (EDSR) [5] also introduced a deeper network with residual connection from ResNet [6]. In more recent years, significant quality improvements have been achieved by several generative adversarial network (GAN)-based SR methods [7]. A super-resolution generative adversarial network (SRGAN) [8] achieved significant improvement in pixel-wise accuracy by introducing a discriminator to their ResNet-like SR network. An enhanced super-resolution generative adversarial network (ESRGAN) [9] made even more improvements with a DenseNet-like generator [10] and Relativistic discriminator [11].

Along with the advancement of super-resolution methods for general images, studies for applying SR for medical images have also been made. Pham et al. [12] applied SRCNN to MR images to enhance spatial resolution. The improvement with GAN-based techniques has also been applied to medical imaging fields [13]. Sánchez and Vilaplana [14] utilized a simplified version of SRGAN to MR images. Yamashita and Markov [15] improved the quality of optical coherence tomography (OCT) images with ESRGAN.

3. Proposed Method

In this paper, we propose a new super-resolution technique for brain MR images with a significantly smaller number of training images. To train the GAN-based super-resolution network, SPS randomly selects many small cubic regions from the input images and feeds them into the network. While the SPS enables the network to be effectively trained with a few images, it also introduces intensity discontinuities around the boundaries of the patches. The ASD suppresses such discontinuities by implicitly knowing the location information of its input patches by referring to both the HR and the generated SR image.

3.1. The Network Architecture

Figure 1 illustrates the schematics of the proposed method. For the network architecture, we used a slightly modified version of the ESRGAN. The modifications we applied are as follows: (1) all the layers such as convolutions, poolings, and pixel-shufflers are changed to their three-dimensional version to process volumetric MR images, and (2) the number of residual-in-residual dense block (RRDB) is reduced from 23 to 5 because the expected input size is smaller than the original ESRGAN.

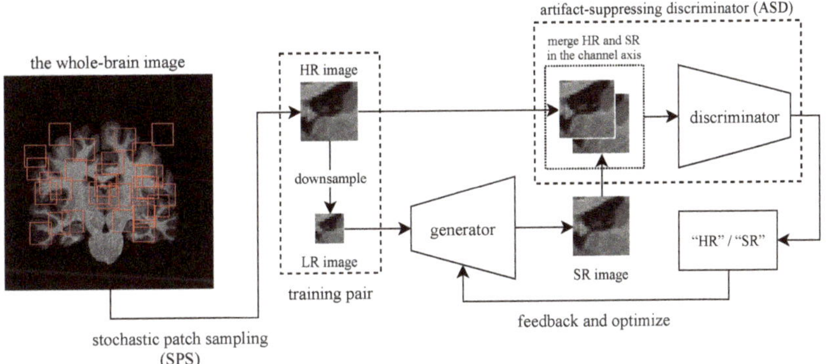

Figure 1. The schematics of the proposed method. Note that all 3D images are drawn in 2D for the sake of visibility.

3.2. Stochastic Patch Sampling (SPS)

Each image is split into a set of smaller three-dimensional cubic patches by randomly choosing the coordinate in the image space. If the patch is sampled from the background and does not contain any brain structure, the patch is automatically rejected and repeatedly re-sampled until sampled from the appropriate coordinate. While the total amount of information fed to the network is theoretically identical, using a collection of sampled patches has several benefits rather than using the whole image at once. Compared to the whole-brain image, the sampled patch is relatively tiny. With this approach, we can use more training samples per batch during mini-batch learning, enabling more stable optimization. The network will also be more robust for unregistered images since the input patches have more significant spatial variances introduced in random sampling. At the inference phase, patches are sampled evenly from the input image with the grid manner. Then, each patch is upscaled by the network and combined into a single image.

3.3. Artifact Suppressing Discriminator (ASD)

Since the network processes small patches separately and performs super-resolution on each one individually, there is no mechanism to sustain the consistency of the final combined image. This lack of consistency causes discontinuities at each patch's joints, resulting in an unpleasant final image. To address this issue, we introduced ASD, which is an extension of the common GAN discriminator. ASD takes two images; one is always a "real" (or HR, in the context of super-resolution) image, and the other is a generated image or an HR image combined as a two-channel image. Accordingly, the discriminator takes (HR+HR) or (HR+SR) images during training. The proposed ASD can extract more discriminative feature representations by learning the correlation/difference between HR and SR images, while common discriminators take HR and SR images independently.

4. Experiments

The purpose of super-resolution is to aid clinicians or computers in analyzing images more precisely, providing more information on smaller structures. To confirm the effectiveness of the proposed techniques, we investigate the impact of the proposed super-resolution network on disease classification performance in addition to regular image quality evaluation.

4.1. Dataset and Preprocessing

As the high-resolution reference images, we used 37 scans of T1-weighted MR images from the DS000113 ("Forrest Gump") dataset [16] and 11 images from the DS002702 dataset [17], both published by OpenNeuro (https://openneuro.org/, accessed on 20 September 2020). Both datasets are provided as a collection of functional-MRI (fMRI) images but also contain T1-weighted still MR images we used, which were taken on a high-field 7T scanners. After the skull removal and intensity normalization, each HR image is shrunken to 50% to make an LR image, providing high- and low-resolution training pairs. At the SPS phase during the training, we randomly sampled 2500 $24 \times 24 \times 24$ patches per one high-resolution image and downsampled them to 50% resolution to make low-resolution images.

4.2. Training of the Network

Since the network input and output are three-dimensional volume data, we cannot use the perceptual loss with a VGG network used in the original SRGAN and ESRGAN because it must be pre-trained with the ImageNet dataset. To train the network to generate images with more fidelity, we added mean-squared error (MSE) of gradients of the images for all directions to capture finer transitions of the intensity.

As for optimization of both networks, we used the Adam optimizer with the same learning rate and β_1, β_2 parameters with the original ESRGAN.

4.3. Assessing the Image Quality

First, we measure the two most standard metrics for assessing a super-resolution system: (1) peak signal-to-noise ratio (PSNR) and (2) structural similarity (SSIM) between each output image and its corresponding original high-resolution image. In general, SR studies using GAN, Inception Score, and Fréchet Inception Distance (FID) are often used. However, scores are calculated based on low-dimensional representations of two-dimensional images by models trained on everyday objects (e.g., ImageNet) and are unsuitable for this evaluation. We also investigate the line profile of the optic thalamus, which is difficult to see for fine structure with a conventional MR scanner.

4.4. Assessing the Impact on Improving Diagnostic Performance

In addition, the purpose of super-resolution is to aid clinicians or computers in analyzing images more precisely, providing more information on smaller structures. Therefore, we investigate the effectiveness of the proposed SR method on disease classification performance. This way, we can emulate one of the real-world applications of super-resolution for medical images.

In this experiment, we used 650 images from the ADNI2. (Data used in the preparation of this article were obtained from the Alzheimer's Disease Neuroimaging Initiative (ADNI) database (adni.loni.usc.edu). The ADNI was launched in 2003 as a public-private partnership, led by Principal Investigator Michael W. Weiner, MD. The primary goal of ADNI has been to test whether serial magnetic resonance imaging (MRI), positron emission tomography (PET), other biological markers, and clinical and neuropsychological assessment can be combined to measure the progression of mild cognitive impairment (MCI) and early Alzheimer's disease (AD). For up-to-date information, see www.adni-info.org, accessed on 20 September 2020) dataset, containing 360 cognitively normal (CN) images and 290 Alzheimer's disease (AD) images. Each image is downsampled into 1.4-mm pixel spacing to match the training data after skull removal and intensity normalization. We first performed super-resolution for all images to make pseudo-high-resolution training/validation samples. Then, we trained a three-dimensional version of MobileNetV2 with 90% of the images and evaluated the area-under-curve (AUC) score with the remaining 10% of them.

To confirm that the proposed SR process recovers some of the lost information from low-resolution images, we also trained a classifier with downsampled images. Then, we trained another classifier with super-resolved downsampled images and compared their AUC score. We used 50% and 25% for the downsampling scales and three-dimensional MobileNetV2 for the classifier network.

Here, we defined "recovery ratio" to measure how much information is recovered from a low-resolution image as follows:

$$\text{recovery ratio} = \frac{\text{AUC(SR)} - \text{AUC(LR)}}{\text{AUC(HR)} - \text{AUC(LR)}},$$

where AUC(HR), AUC(LR), and AUC(SR) are the AUC score on HR images, LR ($\times 0.5$ downsampled from HR) images, and SR images, where $2\times$ super-resolved images are applied for LR images, respectively. Here, we assume that the AUC with the SR images does not exceed that with the HR images, i.e., the AUC with the HR dataset is the upper bound for the resolution.

5. Results

5.1. Image Quality Assessment

Figures 2 and 3 show examples of MR images generated by proposed super-resolution networks and their original HR images, and their magnified view, respectively. In Table 1, the average SSIM and PSNR between super-resolved images using each method and their ground-truth high-resolution images are also summarized.

The output images of the network without SPS, i.e., plain ESRGAN (column (2)), are visibly blurry, and most of the structural features are lost, leading to the lower SSIM/PSNR value. With the proposed SPS (column (3)), generated images are significantly sharper and visibly natural-looking. However, grid-shaped intensity shifts appear at the joints of each patch (Figure 3 (3)). On the other hand, almost all the intensity shifts are suppressed with the images with the proposed discriminator (column (4)).

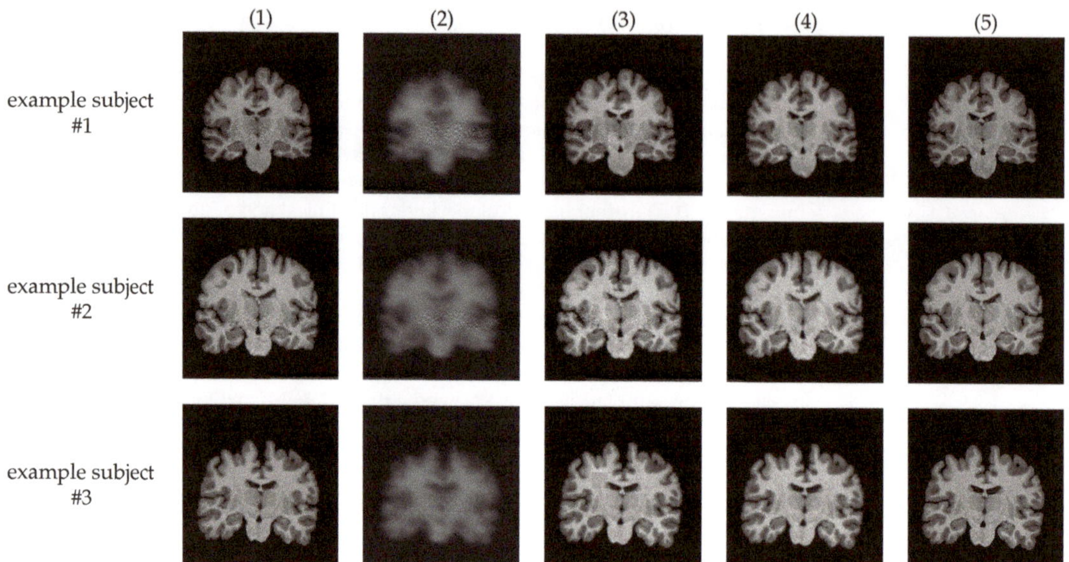

Figure 2. Examples of generated images with their input and ground-truth images. (1) low-resolution input image, (2) ESRGAN without SPS and ASD, (3) ESRGAN + SPS, **(4) ESRGAN + SPS + ASD (proposed)**, and (5) ground truth high-resolution image.

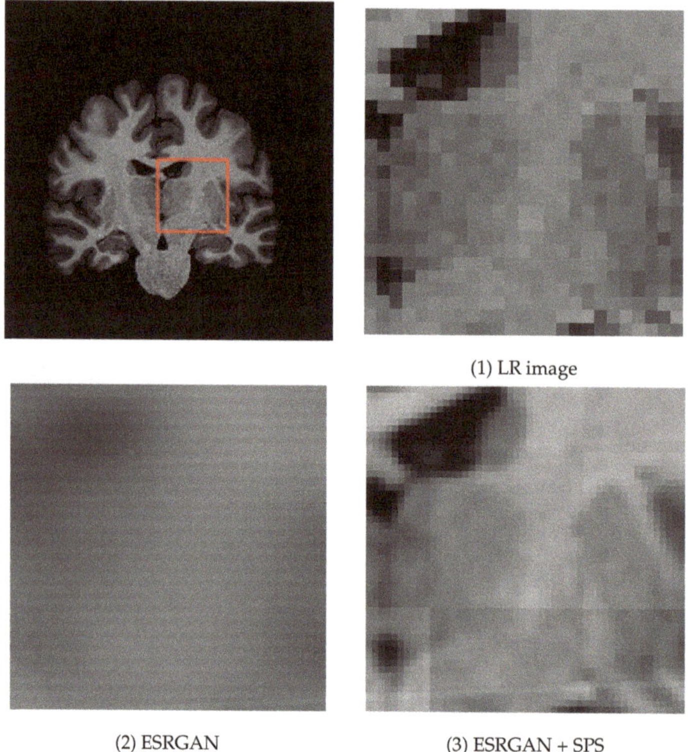

Figure 3. *Cont.*

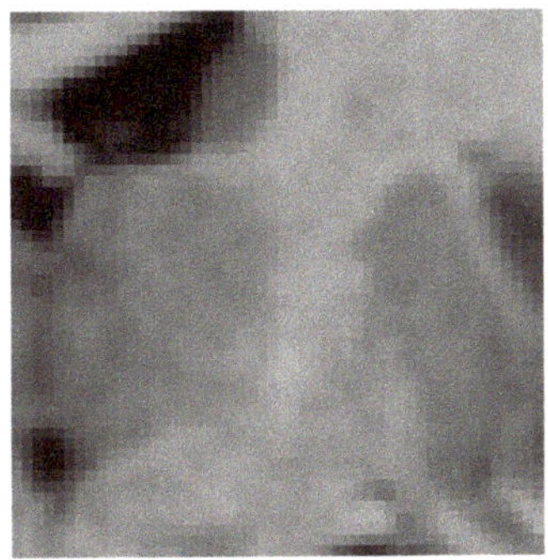

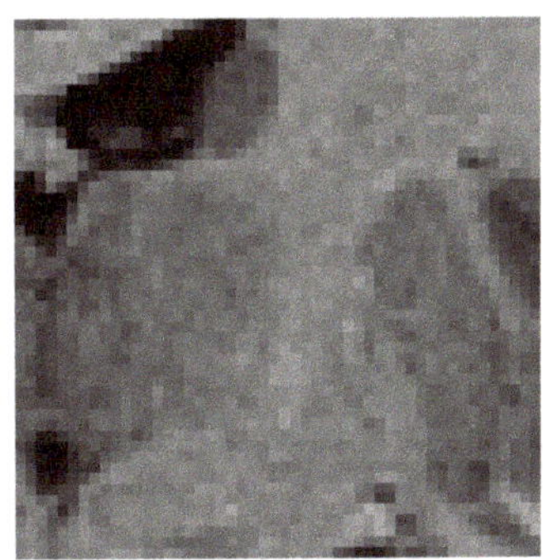

(4) ESRGAN + SPS + ASD (proposed) (5) ground truth high-resolution image

Figure 3. Magnified example of the the generated image.

Table 1. Values of image quality measurements with each method.

SR Method	SSIM	PSNR
ESRGAN	0.7659	16.19
ESRGAN + SPS	0.9056	25.81
ESRGAN + SPS + ASD (proposed)	**0.9443**	**26.92**

5.2. Effectiveness on Classification Performance

Table 2 shows the diagnostic performances of the Alzheimer's disease classifier trained and tested with the super-resolved images. The performance of pure ESRGAN is abysmal because it fails to generate images, and the proposed method outperforms the other methods.

Table 2. AUC scores on Alzheimer's disease diagnosis of the networks trained with images generated by each SR method.

SR Method as Preprocessing	AUC
None	76.17
ESRGAN	67.35
ESRGAN + SPS	79.01
ESRGAN + SPS + ASD (proposed)	**81.57**

Table 3 and Figure 4 summarize the classification accuracies with the downsampled images with/without super-resolution. For the sake of better comparison, results on original (non-downsampled) images are also listed/plotted in them.

Table 3. AUC scores on Alzheimer's disease diagnosis of the networks trained with downsampled images.

Image Scale	With SR (%)	Without SR (%)	Recovery Ratio
1×	(a) 79.01	(d) 76.17	N/A
0.5×	(b) 71.37	(e) 70.46	15.94
0.25×	(c) 60.87	(f) 52.86	45.51

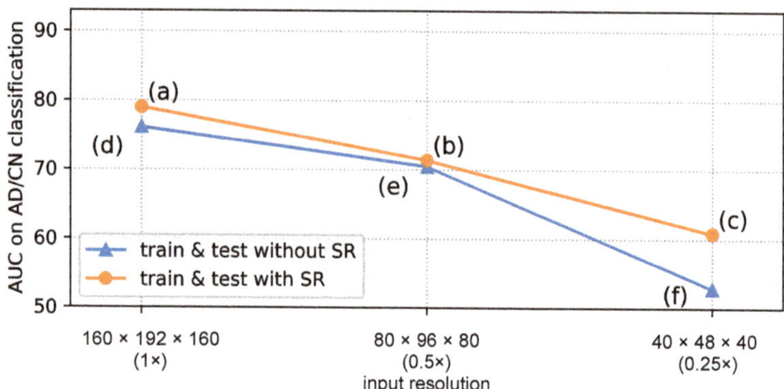

Figure 4. AUC score comparison of networks, trained with downsampled images.

6. Discussion

6.1. Qualities of Generated SR Images

ESRGAN, a sophisticated super-resolution method based on GAN that requires many training images, could not generate any images at all with about 30 training images. This result is worse than the result from the bi-cubic interpolation of LR images. By introducing patch learning with the proposed SPS, we can confirm that it is possible to generate images with a certain level of accuracy. However, as mentioned earlier, discontinuities between patches are noticeable.

With the introduction of the proposed ASD, the discontinuities are mostly suppressed and achieved to generate more natural-looking super-resolution images. Besides the quantitative metrics such as PSNR and SSIM, the line-profile shown in Figure 5 shows that the proposed method can generate the details of finer tissues, which are known to be more challenging to capture with conventional MRI scanners.

In regular GAN training, a generator and a discriminator are trained adversarially. The discriminator indirectly lets the generator learn to make more natural images by trying to identify whether the input is "real" or "fake" (i.e., HR or SR). On the other hand, the proposed ASD takes a two-channeled input, which always contains a HR image in one of the channels. Therefore, in addition to the usual effect, the discriminator itself learns the SR image closer to the HR by implicitly giving the information of the relative location of a patch in a whole-brain image. In this regard, an overfitting effect could be expected because of more information given to the networks during training. Nevertheless, from our experiment with different patients, no adverse effect was confirmed.

6.2. Impact of Super-Resolution on the Disease Classification Problem

The removal of unwanted boundary discontinuities by ASD resulted in an improved AUC score by 5.4% in diagnosing Alzheimer's disease. The increased visibility of essential structures, as shown in Figure 5, is thought to have contributed to the improved diagnostic performance.

In the experiment with downsampled images, first, we confirmed that the AUC score drops as the input resolution decreases, as we intuitively expected. From Table 3, it can be said that the images enhanced by the proposed method can boost the performance up to 45% closer to the possible upper bound of score.

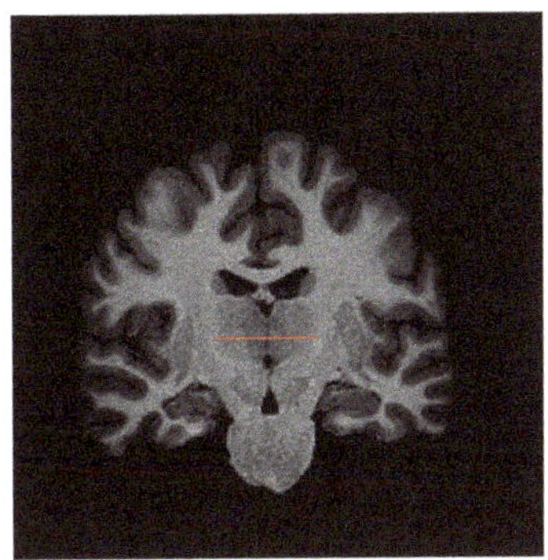

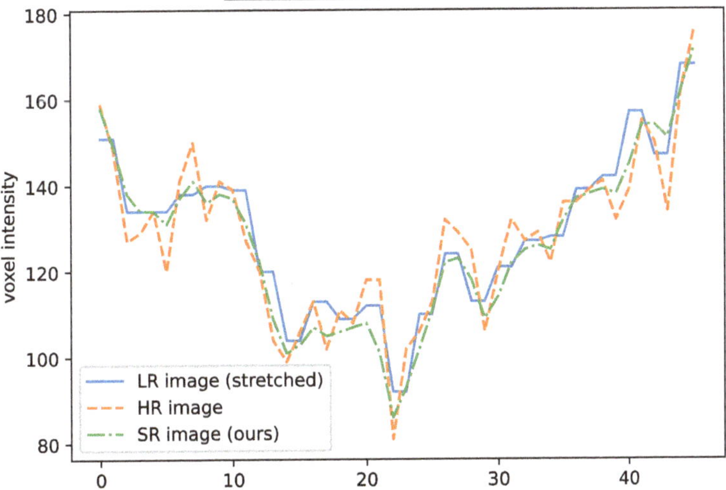

Figure 5. Line profile of the thalamus. The thalamus contains subnuclei with characteristic signal intensity, but it is challenging to identify thalamic subnuclei because low-resolution MRI does not provide sufficient contrast. Our SR method can obtain intrathalamic contrast equivalent to that of HR images.

6.3. Limitations and Future Work

In the proposed method, a low-resolution training image is obtained by simply downsampling the corresponding high-resolution image. However, the actual differences between images with high-field scanners and ordinary ones are not just image resolution but also intensity contrasts, the amount of noise, and so on. The network would perform better if we trained it with high-field and actual ordinary scanners. In the future, we will use more HR images to develop a better method.

7. Conclusions

In this paper, we propose a new super-resolution method for brain MR images with a significantly smaller number of training images. Our method is GAN-based super-resolution with two essential proposed techniques: the SPS and ASD. These proposed techniques succeeded in generating super-resolution images from the training of only about 30 brain MR images. The images generated in this way showed an overall improvement in image quality and an increase in the resolution of critical diagnostic regions, which helped to improve the disease diagnostic performance of the CNN-based classifier built on these images.

Author Contributions: Conceptualization, methodology, K.I., H.I.; software, K.I.; validation, K.I., H.I., K.O.; resources, data curation, K.O.; writing—original draft preparation, K.I.; writing—review and editing, K.I., H.I., K.O.; supervision, H.I., K.O.; project administration, H.I. All authors have read and agreed to the published version of the manuscript.

Funding: This research was supported in part by the Ministry of Education, Science, Sports and Culture of Japan (JSPS KAKENHI), Grant-in-Aid for Scientific Research (C), 21K12656, 2021–2023.

Institutional Review Board Statement: Not applicable.

Informed Consent Statement: Informed consent was obtained from all subjects involved in the study.

Data Availability Statement: We used a brain MR image dataset published by the OpenNeuro and Alzheimer's Disease Neuroimaging Initiative (ADNI) project.

Acknowledgments: The MRI data collection and sharing for this project was funded by the Alzheimer's Disease Neuroimaging Initiative (ADNI) (National Institutes of Health Grant U01 AG024904) and DOD ADNI (Department of Defense award number W81XWH-12-2-0012). ADNI is funded by the National Institute on Aging, the National Institute of Biomedical Imaging and Bioengineering, and through generous contributions from the following: AbbVie, Alzheimer's Association; Alzheimer's Drug Discovery Foundation; Araclon Biotech; BioClinica, Inc.; Biogen; Bristol-Myers Squibb Company; CereSpir, Inc.; Cogstate; Eisai Inc.; Elan Pharmaceuticals, Inc.; Eli Lilly and Company; EuroImmun; F. Hoffmann-La Roche Ltd and its affiliated company Genentech, Inc.; Fujirebio; GE Healthcare; IXICO Ltd.; Janssen Alzheimer Immunotherapy Research & Development, LLC.; Johnson & Johnson Pharmaceutical Research & Development LLC.; Lumosity; Lundbeck; Merck & Co., Inc.; Meso Scale Diagnostics, LLC.; NeuroRx Research; Neurotrack Technologies; Novartis Pharmaceuticals Corporation; Pfizer Inc.; Piramal Imaging; Servier; Takeda Pharmaceutical Company; and Transition Therapeutics. The Canadian Institutes of Health Research is providing funds to support ADNI clinical sites in Canada. Private sector contributions are facilitated by the Foundation for the National Institutes of Health (www.fnih.org). The grantee organization is the Northern California Institute for Research and Education, and the study is coordinated by the Alzheimer's Therapeutic Research Institute at the University of Southern California. ADNI data are disseminated by the Laboratory for Neuro Imaging at the University of Southern California.

Conflicts of Interest: The authors declare no conflict of interest.

References

1. Greenspan, H. Super-resolution in Medical Imaging. *Comput. J.* **2009**, *52*, 43–63. [CrossRef]
2. Wang, Z.; Chen, J.; Hoi, S.C. Deep Learning for Image Super-resolution: A Survey. *IEEE Trans. Pattern Anal. Mach. Intell.* **2020**. [CrossRef] [PubMed]
3. Dong, C.; Loy, C.C.; He, K.; Tang, X. Image Super-resolution using Deep Convolutional Networks. *IEEE Trans. Pattern Anal. Mach. Intell.* **2015**, *38*, 295–307. [CrossRef] [PubMed]
4. Kim, J.; Lee, J.K.; Lee, K.M. Accurate image super-resolution using very deep convolutional networks. In Proceedings of the IEEE Conference on Computer Vision and Pattern Recognition, Las Vegas, NV, USA, 27–30 June 2016; pp. 1646–1654.
5. Lim, B.; Son, S.; Kim, H.; Nah, S.; Mu Lee, K. Enhanced Deep Residual Networks for Single Image Super-resolution. In Proceedings of the IEEE Conference on Computer Vision and Pattern Recognition Workshops, Honolulu, HI, USA, 26 July 2017; pp. 136–144.
6. He, K.; Zhang, X.; Ren, S.; Sun, J. Deep Residual Learning for Image Recognition. In Proceedings of the IEEE Conference on Computer Vision and Pattern Recognition, Las Vegas, NV, USA, 27–30 June 2016; pp. 770–778.
7. Goodfellow, I.; Pouget-Abadie, J.; Mirza, M.; Xu, B.; Warde-Farley, D.; Ozair, S.; Courville, A.; Bengio, Y. Generative Adversarial Nets. *Adv. Neural Inf. Process. Syst.* **2014**, *27*, arXiv:1406.2661.
8. Ledig, C.; Theis, L.; Huszár, F.; Caballero, J.; Cunningham, A.; Acosta, A.; Aitken, A.; Tejani, A.; Totz, J.; Wang, Z.; et al. Photo-realistic Single Image Super-resolution using a Generative Adversarial Network. In Proceedings of the IEEE Conference on Computer Vision and Pattern Recognition, Honolulu, HI, USA, 26 July 2017; pp. 4681–4690.
9. Wang, X.; Yu, K.; Wu, S.; Gu, J.; Liu, Y.; Dong, C.; Qiao, Y.; Change Loy, C. ESRGAN: Enhanced Super-resolution Generative Adversarial Networks. In Proceedings of the European Conference on Computer Vision (ECCV) Workshops, Munich, Germany, 8–14 September 2018.
10. Huang, G.; Liu, Z.; Van Der Maaten, L.; Weinberger, K.Q. Densely Connected Convolutional Networks. In Proceedings of the IEEE Conference on Computer Vision and Pattern Recognition, Honolulu, HI, USA, 21–26 July 2017; pp. 4700–4708.
11. Jolicoeur-Martineau, A. The Relativistic Discriminator: A Key Element Missing From Standard GAN. *arXiv* **2018**, arXiv:1807.00734.

12. Pham, C.H.; Ducournau, A.; Fablet, R.; Rousseau, F. Brain MRI Super-resolution Using Deep 3D Convolutional Networks. In Proceedings of the 2017 IEEE 14th International Symposium on Biomedical Imaging (ISBI 2017), Melbourne, Australia, 18–21 April 2017; pp. 197–200.
13. Yi, X.; Walia, E.; Babyn, P. Generative Adversarial Network in Medical Imaging: A Review. *Med. Image Anal.* **2019**, *58*, 101552. [CrossRef]
14. Sánchez, I.; Vilaplana, V. Brain MRI Super-resolution Using 3D Generative Adversarial Networks. *arXiv* **2018**, arXiv:1812.11440.
15. Yamashita, K.; Markov, K. Medical Image Enhancement Using Super Resolution Methods. In Proceedings of the International Conference on Computational Science, Las Vegas, NV, USA, 16–18 December 2020; pp. 496–508.
16. Hanke, M.; Baumgartner, F.J.; Ibe, P.; Kaule, F.R.; Pollmann, S.; Speck, O.; Zinke, W.; Stadler, J. Forrest Gump. OpenNeuro. 2018. Available online: https://openneuro.org/datasets/ds000113/versions/1.3.0 (accessed on 20 September 2020). [CrossRef]
17. Kay, K.; Jamison, K.W.; Vizioli, L.; Zhang, R.; Margalit, E.; Ugurbil, K. High-Field 7T Visual fMRI Datasets. OpenNeuro. 2020; Available online: https://openneuro.org/datasets/ds002702/versions/1.0.1 (accessed on 20 September 2020). [CrossRef]

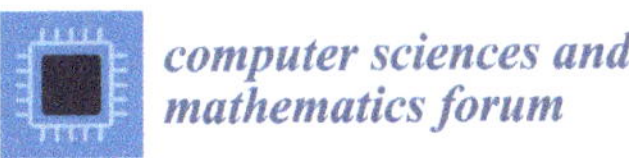

Proceeding Paper

Dual Complementary Prototype Learning for Few-Shot Segmentation †

Qian Ren and Jie Chen *

School of Electronic and Computer Engineering, Peking University, Shenzhen 518055, China; renq2019@pku.edu.cn
* Correspondence: chenj@pcl.ac.cn

† Presented at the AAAI Workshop on Artificial Intelligence with Biased or Scarce Data (AIBSD), Online, 28 February 2022.

Abstract: Few-shot semantic segmentation aims to transfer knowledge from base classes with sufficient data to represent novel classes with limited few-shot samples. Recent methods follow a metric learning framework with prototypes for foreground representation. However, they still face the challenge of segmentation of novel classes due to inadequate representation of foreground and lack of discriminability between foreground and background. To address this problem, we propose the Dual Complementary prototype Network (DCNet). Firstly, we design a training-free Complementary Prototype Generation (CPG) module to extract comprehensive information from the mask region in the support image. Secondly, we design a Background Guided Learning (BGL) as a complementary branch of the foreground segmentation branch, which enlarges difference between the foreground and its corresponding background so that the representation of novel class in the foreground could be more discriminative. Extensive experiments on PASCAL-5^i and COCO-20^i demonstrate that our DCNet achieves state-of-the-art results.

Keywords: few-shot; semantic segmentation

Citation: Ren, Q.; Chen, J. Dual Complementary Prototype Learning for Few-Shot Segmentation. *CSFM* **2022**, 3, 8. https://doi.org/10.3390/cmsf2022003008

Academic Editors: Kuan-Chuan Peng and Ziyan Wu

Published: 29 April 2022

1. Introduction

Attributed to the development of convolutional neural networks (CNNs) with its strong representation ability and the access of large-scale datasets, semantic segmentation and object detection have developed tremendously. However, it is worth to point out that annotating a large number of object masks is time-consuming, expensive, and sometimes infeasible in some scenarios, such as computer-aided diagnosis systems. Moreover, without massive annotated data, the performance of deep learning models drops dramatically on classes that do not appear in the training dataset. Few-shot segmentation (FSS) is a promising field to tackle this issue. Unlike conventional semantic segmentation, which merely segments the classes appearing in the training set, few-shot segmentation utilizes one or a few annotated samples to segment new classes.

They firstly extract features from both query and support images, and then the support features and their masks are encoded into a single prototype [1] to represent foreground semantics or a pair of prototypes [2,3] to represent the foreground and background. Finally, they conduct dense comparison between prototype(s) and query feature. Feature comparison methods are usually performed in one of two ways: explicit metric function, (e.g., cosine-similarity [3]) and implicit metric function (e.g., relationNet [4]).

As shown in Figure 1a, it is common-sense [2,5,6] that using a single prototype generated by masked average pooling is unable to carry sufficient information. Specifically, due to variant appearance and poses, using masked average pooling only retains the information of discriminative pixels and ignores the information of plain pixels. To overcome this problem, multi-prototype strategy [2,5,6] is proposed by dividing foreground regions into several pieces.

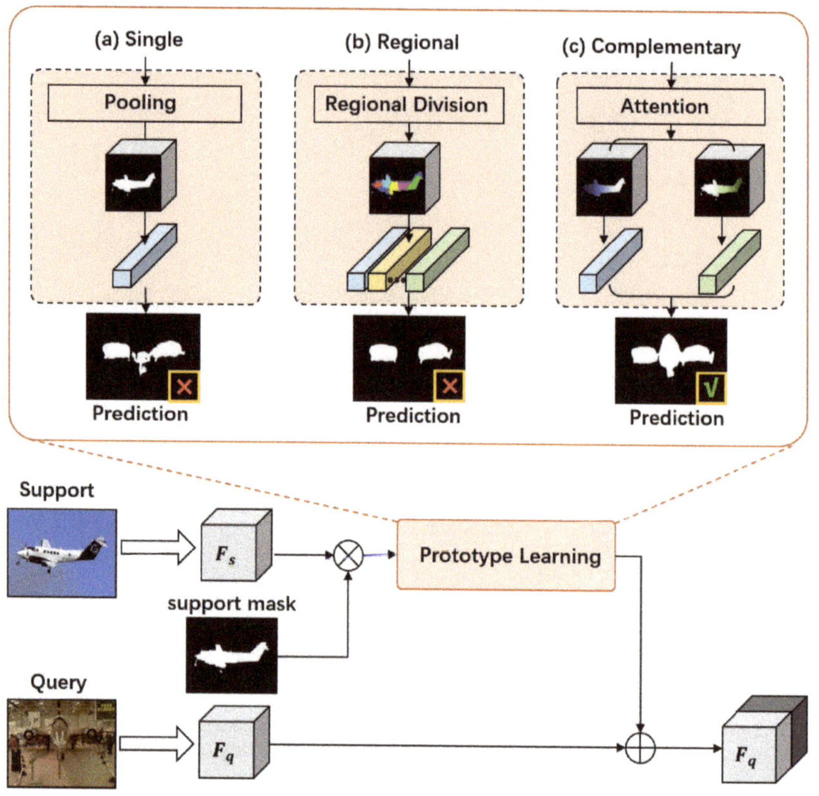

Figure 1. Illustration of difference in prototype learning for 1-shot segmentation. (**a**) Single prototype methods [1,7] tend to lose information as plain pixels. (**b**) Multi-prototype methods [2,5,8] based on regional division may damage the representation for the whole object. (**c**) Our Complementary Prototype Generation module retains the information of discriminative pixels and plain pixels adaptively.

However, as shown in Figure 1b, these multi-prototype methods still suffer from two drawbacks. Firstly, the whole representation of foreground region is weakened, since existing methods split regions into several pieces and damage the correlation among the generated prototypes. Moreover, current methods often ignore inter-class similarity between foreground and background, and their training strategy in the context of segmenting the main foreground objects leads to underestimating the discrimination between the foreground and background. As a result, existing multi-prototype methods tend to misclassify background pixels into foreground.

In this paper, we propose a simple yet effective method, called Dual Complementary prototype Network (DCNet), to overcome the above mentioned drawbacks. Specifically, it is composed of two branches to segment the foreground and background in a complementary manner, and both segmentation branches rely on our proposed Complementary Prototype Generation (CPG) module. The CPG module is proposed to extract comprehensive support information from the support set. Through global average pooling with support mask, we extract the average prototype at first, and we obtain its attention weight on the support image by calculating the cosine distance between the foreground feature and the average prototype iteratively. In this way, we can easily figure out which part of the information is focused and which part of the information is ignored without segmentation on support image. Then we use this attention weight to generate a pair of prototypes to represent the focused and the ignored region. By using a weight map to generate prototypes for comparison, we can preserve the correlation among the generated prototypes and avoid the information loss to a certain extent.

Furthermore, we introduce background guided learning to pay additional attention on the inter-class similarity between the foreground and background. Considering that the background in support images is not always the same as that in a query image, we adopt a different training manner from foreground segmentation, where the query background mask is used as guidance for query image background segmentation. In this way, our model could learn a more discriminative representation for distinguishing foreground and background. The proposed method effectively and efficiently improves the performance on FSS benchmarks without extra inference cost.

The main contributions of this work are summarized as follows.

1. We propose Complementary Prototype Generation (CPG) to learn powerful prototype representation without extra parameters costs;
2. We propose Background Guided Learning (BGL) to increase the feature discrimination between foreground and background. Besides, BGL is merely applied in the training phase so that it would not increase the inference time;
3. Our approach achieves the state-of-the-art results on both PASCAL-5^i and COCO-20^i datasets and improves the performance of the baseline model by 9.1% and 12.6% for 1-shot and 5-shot setting on COCO-20^i.

2. Related Work

2.1. Semantic Segmentation

Semantic segmentation, which aims to perform classification for each pixel, has been extensively investigated. Following Fully Convolution Network (FCN) [9], which uses fully convolutional layers instead of fully connected layers as a classifier for semantic segmentation, large numbers of network frameworks have been designed. For example, Unet adopted a multi-scale strategy and a encoder-decoder architecture to improve the performance of FCN, and PSPNet was proposed to use the pyramid pooling module (PPM) to generate object details. Deeplab [10,11] designed an Atrous Spatial Pyramid Pooling (ASPP) module, conditional random field (CRF) module, and dilated convolution to FCN architecture. Recently, attention mechanism has been introduced, PSANet [12] was proposed to use point-wise spatial attention with a bi-directional information propagation paradigm. Channel-wise attention [13] and non-local attention [14–17] are also effective for segmentation. These methods have managed to succeed in large-scale datasets but they are not designed to deal with rare and unseen classes and cannot be accommodated without fine-tuning.

2.2. Few-Shot Learning

Few-shot learning focuses on the generalization ability of models, so that they can learn to predict novel classes with a few annotated examples [4,18–21]. Matching networks [19] were proposed for 1-shot learning to exploit a special kind of mini-batches called episodes to match the training and testing environments, enhancing the generalization on the novel classes. Prototypical network [20] was introduced to compute the distances between the representation cluster centers for few-shot classification. Finn et al. [21] proposed an algorithm for meta-learning that is model-agnostic. Even though few shot learning has been extensively studied for classification task, it is still hard to adopt few-shot learning directly on segmentation due to the dense prediction.

2.3. Few-Shot Segmentation

As the extension of few-shot learning, few-shot semantic segmentation has also received considerable attention very recently. Shaban et al. first proposed the few-shot segmentation problem with a two-branch conditional network that learned the parameters on support images. Different from [22], later works [1–3,23,24] follow the idea of metric learning. Zhang et al. generates the foreground object segmentation of the support class by measuring the embedding similarity between query and supports, where their embeddings are extracted by the same backbone model. Generally, metric learning based methods can be divided into two groups: one group is inspired by ProtoNet [20], e.g., PANet [3] first embeds different foreground objects and the background into different prototypes via a shared feature extractor, and then measures the similarity between the query and the prototypes. The other group is inspired by relationNet [4], which learns a metric function to measure the similarity, e.g., Refs. [1,7,8] use an FPN-like structure to perform dense comparison with affinity alignment. Then, considering the incomplete representation of a single prototype, Li et al. [5] divide the masked region into pieces, the number of which is decided by the area of the masked region and then conducts masked average pooling for each piece to generate the numbers of the prototypes.Zhang et al. [6] utilize the uncovered foreground region and covered foreground region through segmentation on support images to generate a pair of prototypes to retrieve the loss information. However, compared to self-segmentation mechanism [6], our CPG does not need to segment on support images and utilization of CPG obtains competitive performance with few cost.s Compared to cluster methods [5,8], the experiment in the ablation study shows that our method can avoid over-fitting and generate stable performance in each setting.

Moreover, recent methods such as MLC [25] and SCNet [26] start to make use of knowledge hidden in the background. By exploiting the pre-training knowledge for the discovery of the latent novel class in the background, their methods bring huge improvements to the few-shot segmentation task. However, we argue that such a method is difficult to apply in realistic scenarios, since a novel class object is not only unlabelled but also unseen in the training set. Instead, we propose background guided learning to enhance the feature discriminability between the foreground and the background, which also improves the performance of the model.

3. Proposed Methods

3.1. Problem Setting

The aim of few-shot segmentation is to obtain a model that can learn to perform segmentation from only a few annotated support images in novel classes. The few-shot segmentation model should be trained on a dataset D_{train} and evaluated on a dataset D_{test}. Given the classes set in D_{train} is C_{train} and classes set in D_{test} is C_{test}, there is no overlap between training classes and test classes, e.g., $C_{train} \cap C_{test} = \emptyset$.

Following a previous definition [22], we divide the images into two non-overlapping sets of classes C_{train} and C_{test}. The training set D_{train} is built on C_{train} and the test set is built on C_{test}. We adopt the episode training strategy, which has been demonstrated as an effective approach for few-shot recognition. Each episode is composed of a shot support set $S = \{I_k^s, M_k^s\}_{k=1}^K$ and a query set $Q = I^q, M^q$ to form a K-shot episode $\{S, I^q\}$, where I^* andM^* are the image and its corresponding mask label, respectively. Then, the training set and test set are denoted by $D_{train} = \{S\}^{N_{train}}$ and $D_{test} = \{Q\}^{N_{test}}$, where N_{train} and N_{test} is the number of episodes for the training and test set. Note that both the mask M^s of the support set and the mask M^q of the query set are provided in the training phase, but only the support image mask M^s is included in the test phase.

3.2. Overview

As shown in Figure 2, our Dual Complementary prototype Network (DCNet) is trained via the episodic scheme on the support-query pairs. In episodic training, supports images and a query image are input to the share-weight encoder for feature extraction. Then, the query feature is compared with prototypes of the current support class to generate a foreground segmentation mask via a FPN-like decoder. Besides, we propose an auxiliary supervision, named Background Guided Learning (BGL), where our network learns robust prototype representation for a class-agnostic background in an embedding space. In this supervision, the query feature is compared with prototypes of the query background to make a prediction on its own background. With this joint training strategy, our model can learn discriminative representation for foreground and background.

Thus, the overall optimization target can be briefly formulated as:

$$\mathcal{L}_{overall} = \mathcal{L}_{fg} + \gamma\mathcal{L}_{bg}, \tag{1}$$

where $\mathcal{L}_{fg}$ and $\mathcal{L}_{bg}$ denote the foreground segmentation loss and background segmentation loss, respectively, and γ is the balance weight, which is simply set as 1.

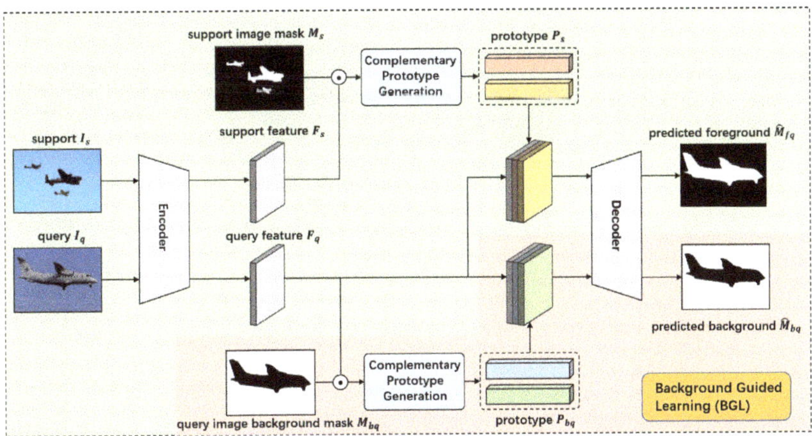

Figure 2. The framework of the proposed DCNet for 1-shot segmentation. At first, the encoder generates feature maps F_s and F_q from the support images and query images. Then, the support image masks M_s and related features are fed into CPG to generate a pair of foreground prototypes P_s. Finally, P_s is expanded and concatenated with the query feature F_q as an input to the decoder to predict the foreground in the query image. In the meantime, in BGL, the query feature F_q and its background mask M_{bq} are fed into CPG to generate a pair of background prototypes P_{bq}. P_{bq} is expanded and concatenated with query feature F_q as an input to the decoder to predict the background in the query image.

In the following subsections, we first elaborate our prototype generation algorithm. Then, background-guided learning on 1-shot setting is introduced, followed by inference.

3.3. Complementary Prototypes Generation

Inspired by SCL [6], we propose a simple and effective algorithm, named Complementary Prototypes Generation (CPG), as shown in Figure 3. This CPG algorithm generates a pair of complementary prototypes and aggregates information hidden in features based on cosine similarity. Specifically, given the support feature $F \in \mathbb{R}^{H\times W\times C}$ with the mask region as $M \in \mathbb{R}^{H\times W}$, we extract a pair of prototypes to fully represent the information in the mask region.

As the first step, we extract the targeted feature $F' \in \mathbb{R}^{H\times W\times C}$ filtered through mask M from F, in Equation (2),

$$F' = F \odot M \tag{2}$$

where $\odot$ represents element-wise multiplication. Then, we initiate prototype P_0 by masked average pooling, in Equation (3),

$$P_0 = \frac{\sum_i^H \sum_j^W F'_{i,j}}{\sum_i^H \sum_j^W M_{i,j}} \tag{3}$$

where i, j represents the coordination of each pixel, H, W denotes the width and height of feature F', respectively. Since $M_{i,j} \in 0, 1$, the sum of M represents the area of the foreground region.

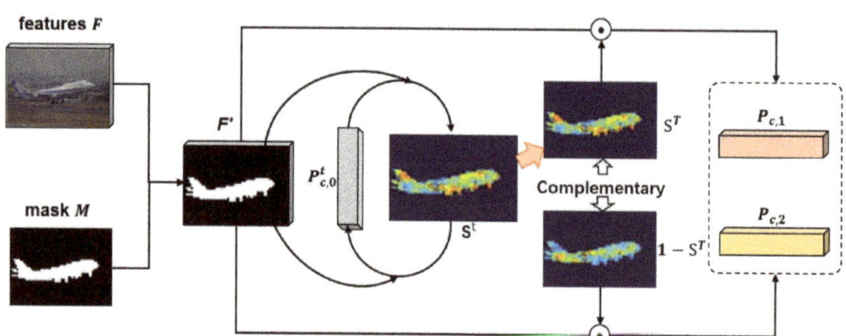

Figure 3. Illustration of the proposed Complementary Prototypes Generation. Similarity S^t and prototype $P^t_{c,0}$ is obtained in t-th iteration. The red arrow indicates the final result S^T after T iterations.

In the next step, we aggregate the foreground features into two complementary clusters. For each iteration t, we first compute the cosine distance matrix $S^t \in \mathbb{R}^{H \times W}$ between the prototype P_0^{t-1} and the targeted features F' as follows,

$$S^t = \text{cosine}(F', P_0^{t-1}) \tag{4}$$

As we keep the relu layer in the encoder layer, the cosine distance is limited in $[0, 1]$. To calculate the weight of target features contributed to P_0^t, we normalize the S matrix as:

$$S^t_{i,j} = \frac{S^t_{i,j}}{\sum_i^H \sum_j^W S^t_{i,j}} \tag{5}$$

Then, after the end of the iteration, based on matrix S^t, we aggregate the features into two complementary prototypes as:

$$P_1 = \sum_i^H \sum_j^W S_{i,j} * F'_{i,j} \tag{6}$$

$$P_2 = \sum_i^H \sum_j^W (1 - S_{i,j}) * F'_{i,j} \tag{7}$$

It is worth noting that these prototypes are not separated like priors and CPG algorithm utilizes a weighted map to generate a pair of complementary prototypes. In this way, we retain the correlation between the prototypes. The whole CPG is delineated in Algorithm 1.

Algorithm 1 Complementary Prototypes Generation (CPG).

Input: targeted feature F', corresponding mask M, the number of iteration T.

init prototype $P^0_{c,0}$ by masked average pooling with F'. $P_0 = \frac{\sum_i^H \sum_j^W F'_{i,j}}{\sum_i^H \sum_j^W M_{i,j}}$

for iteration t in {1, .., T} **do**

Compute association matrix S between targeted feature F' and prototype P_0^{t-1}, $S^t = \text{cosine}(F', P_{c,0}^{t-1})$

Standardize association S^t,

$S^t_{i,j} = S^t_{i,j} / (\sum_i^H \sum_j^W S^t_{i,j})$

Update prototype $P_{c,0}$,

$P^t_0 = \sum_i^H \sum_j^W S^t_{i,j} * F'_{i,j}$

end for

generate complementary prototypes P_c from S^T,

$P_1 = \sum_i^H \sum_j^W (S^T_{i,j}) * F'_{i,j}$

$P_2 = \sum_i^H \sum_j^W (1 - S^T_{i,j}) * F'_{i,j}$

return final prototypes P_1, P_2

3.4. Background Guided Learning

In previous works [1,5,6], the background information has not been adequately exploited for few-shot learning. Especially, these methods only use foreground prototypes to make a final prediction on the query image in the training. As a result, the representation on class-agnostic background is the lack of discriminability. To solve this problem, Background Guided Learning (BGL) is proposed via joint training strategy.

As shown in Figure 2, BGL is proposed to segment the background on the query image based on query background mask M_{bq}. As the first step, query feature F_q and its background mask M_{bq} are fed into the CPG module to generate a pair of complementary prototypes $P_{bq} = P_1, P_2$, following Algorithm 1. Next, we concatenate the complementary prototype P_{bq} with all spatial location in query feature map F_q, as Equation (8):

$$F_m = \epsilon(P_1) \oplus \epsilon(P_2) \oplus F_q, \tag{8}$$

where ϵ denotes the expansion operation and $\oplus$ denotes the concatenation operation, P_1 and P_2 are the complementary prototypes P_{bq} as well as F_m, denoting the concatenated feature. Then, concatenate feature F_m is fed into the decoder, generating the final prediction, as shown in Equation (9):

$$\hat{M} = \text{D}(F_m), \tag{9}$$

where $\hat{M}$ is the prediction of the model, D is a decoder. The loss $\mathcal{L}_{bg}$ is computed by:

$$\mathcal{L}_{bg} = \text{CE}(\hat{M}_{bq}, M_{bq}) \tag{10}$$

where $\hat{M}_{bq}$ denotes the background prediction on a query image and CE denotes the cross-entropy loss.

Intuitively, if the model can predict a good segmentation mask for the foreground using a prototype extracted from the foreground mask region, the prototype learned from the background mask region should be able to segment itself well. Thus, our BGL encourages the model to distinguish the background from the foreground better.

3.5. Inference

In the inference phase, we only keep the foreground segmentation branch for the final prediction. For K-shot setting, we following previous works and use the average to generate a pair of complementary prototypes.

4. Experiments

4.1. Dataset and Evaluation Metrics

4.1.1. Datasets

We evaluate our algorithm on two public few-shot datasets: PASCAL-5^i [22] and COCO-20^i [27]. PASCAL-5^i is built from PASCAL VOC 2012 and SBD datasets. COCO-20^i is built from MS-COCO dataset. In PASCAL-5^i, 20 object classes of PASCAL VOC are split into 4 groups, in which each group contains 5 categories. In COCO-20^i, as PASCAL-5^i, we divide MS-COCO into 4 groups, in which each group contains 20 categories. For PASCAL-5^i and COCO-20^i, we evaluate our approach based on PFENet. We use the same categories division and randomly sample 20,000 support-query pairs to evaluate as PFENet.

For both datasets, we adopt 4-fold cross-validation i.e., a training model on three folds (base class) and the inference model on the remaining one (novel class). The experimental results are reported on each test fold, and we also report the average performance of all four test folds.

4.1.2. Evaluation Metrics

Following previous work [7,27], we use the widely adopted class mean intersection over union (mIoU) as our major evaluation metric for the ablation study, since the class mIoU is more reasonable than the foreground-background IoU (FB-IoU), as stated in [7]. For each class, the IoU is calculated by $\frac{TP}{TP+FN+FP}$, where TP denotes the number of true positives, FP denotes the number of false postives and FN denotes the number of false negatives. Then, mIOU is the mean value of all classes IoU in the test set. For FB-IoU, only the foreground and background are considered ($C = 2$). We take the average of the results on all folds as the final mIoU/FB-IoU.

4.2. Implementation Details

Our approach is based on PFENet [1] with ResNet-50 as the backbone to create a fair comparison with the other methods. Following previous work [1,5,6], the parameters of the backbone are initialized with the pre-trained ImageNet, and is kept fixed during training. Other layers are initialized by the default setting of PyTorch. For PASCAL-5^i, the network is trained with an initial learning rate of 2.5×10^{-3}, weight decay of 1×10^{-4}, and a momentum of 0.9 for only 100 epochs. The batch size is 4. For COCO-20^i, the network is trained for 50 epochs with a learning rate of 0.005 and batch size of 8. We use data augmentation during training. Specifically, input images are transformed with random scale, horizontally flipped and rotated from $[-10, 10]$, and then all images are cropped to 473×473 (for PASCAL and COCO) or 641×641 (for COCO) as the training samples, for fair comparison. We implemented our model with 4 RTX2080Ti.

4.3. Comparisons with State-of-the-Art

4.3.1. COCO-20^i Result

COCO-20^i is a very challenging dataset that contains the numbers of objects in realistic scene images. We compare our approach with others on this dataset, and our approach outperforms other approaches by a big margin, as shown in Table 1. It can be seen that our approach achieves state-of-the-art performance on both 1-shot and 5-shot settings with mIOU gain of 0.3% and 0.5%, respectively. Furthermore, compared to our baseline (PFENet with ResNet101), our approach (with ResNet101) obtains 9.1% and 12.6% mIoU increases for 1-shot and 5-shot settings. In Table 2, our method obtains a top-performing 1-shot result and competitive 5-shot result with respect to FB-IoU. Once again, these results demonstrate that the proposed method is able to deal with more complex cases, since MSCOCO is a much more challenging dataset with diverse samples and categories.

Table 1. Comparison with other state-of-the-art methods on COCO-20^i for 1-shot and 5-shot settings. † denotes the model using size 641 × 641 as the training samples. All methods are tested on the original size. **Bold** denotes the best performance and red denotes the second best performance.

Method	Backbone	1-Shot					5-Shot				
		Fold-1	Fold-2	Fold-3	Fold-4	Mean	Fold-1	Fold-2	Fold-3	Fold-4	Mean
PFENet (TPAMI'20)	ResNet101	34.3	33.0	32.3	30.1	32.4	38.5	38.6	38.2	34.3	37.4
SCL (CVPR'21)	ResNet101	36.4	38.6	37.5	35.4	37.0	38.9	40.5	41.5	38.7	39.9
RePRI (CVPR'21)	ResNet101	36.8	41.8	38.7	36.7	38.5	40.4	46.8	43.2	40.5	42.7
FWB (ICCV'19)	ResNet101	17.0	18.0	21.0	28.9	21.2	19.1	21.5	23.9	30.1	23.7
CWT (ICCV'21)	ResNet101	30.3	36.6	30.5	32.2	32.4	38.5	46.7	39.4	43.2	42.0
HSNet (ICCV'21)	ResNet101	37.2	**44.1**	**42.4**	**41.3**	41.2	45.9	**53**	**51.8**	47.1	49.5
SCNet (2021)	ResNet101	38.3	43.1	40.0	39.1	40.1	44.0	47.7	45.0	42.8	44.8
MLC (ICCV'21)	ResNet101	**50.2**	37.8	27.1	30.4	36.4	**57.0**	46.2	37.3	37.2	44.4
SST (IJCAI'20)	ResNet50	-	-	-	-	22.2	-	-	-	-	31.3
DAN (ECCV'20)	ResNet50	-	-	-	-	24.4	-	-	-	-	29.6
PPNet (ECCV'20)	ResNet50	34.5	25.4	24.3	18.6	25.7	48.3	30.9	35.7	30.2	36.2
RPMMs (ECCV'20)	ResNet50	29.5	36.8	28.9	27.0	30.6	33.8	42.0	33.0	33.3	35.5
ASR (CVPR'21)	ResNet50	29.9	35.0	31.9	33.5	32.6	31.3	37.9	33.5	35.2	34.4
ASGNet † (CVPR'21)	ResNet50	-	-	-	-	34.6	-	-	-	-	42.5
CWT (ICCV'21)	ResNet50	32.2	36.0	31.6	31.6	32.9	40.1	43.8	39.0	42.4	41.3
Ours †	ResNet50	37.1	42.8	39.4	37.7	39.3	41.9	49.0	46.3	44.0	45.3
Ours	ResNet101	40.6	**44.1**	40.6	40.2	**41.5**	49.0	52.9	50.5	**47.7**	**50.0**

Table 2. Comparison of FB-IoU on COCO-20^i.

Methods	Backbone	1-Shot	5-Shot
PFENet (TPAMI'20)	ResNet101	58.6	61.9
DAN (ECCV'20)	ResNet101	62.3	63.9
Ours	ResNet101	64.0	68.8

4.3.2. PASCAL-5^i Result

In Table 3, we compare our method with other state-of-the-art methods on PASCAL-5^i. It can be seen that our method achieves on par state-of-the-art performance on 1-shot setting and 5-shot setting. Additionally, our method significantly improves the performance of PFENet on 1-shot and 5-shot segmentation settings, with an mIOU increase of 1.6% and 4%, respectively. In Table 4, our method obtains competitive 1-shot results and top-performing 5-shot results with respect to FB-IoU. In Figure 4, we report some qualitative results generated by our approach with PFENet [1] as the baseline. Our method is capable of making correct predictions and each part of our method could independently improve the performance of the model.

Table 3. Comparison with state-of-the-art methods on PASCAL-5^i for 1-shot and 5-shot settings. For fair comparison, all methods are evaluated with backbone ResNet50 and tested on labels with original sizes. **Bold** denotes the best performance and red denotes the second best performance.

Method	1-Shot					5-Shot				
	Fold-1	Fold-2	Fold-3	Fold-4	Mean	Fold-1	Fold-2	Fold-3	Fold-4	Mean
PGNet (ICCV'19)	56.0	66.9	50.6	50.4	56.0	57.7	68.7	52.9	54.6	58.5
CANet (CVPR'19)	52.5	65.9	51.3	51.9	55.4	55.5	67.8	51.9	53.2	57.1
CRNet (CVPR'20)	-	-	-	-	55.7	-	-	-	-	58.8
SimPropNet (IJCAI'20)	54.9	67.3	54.5	52.0	57.2	57.2	68.5	58.4	56.1	60.0
DAN (ECCV'20)	-	-	-	-	57.1	-	-	-	-	59.5
PPNet (ECCV'20)	47.8	58.8	53.8	45.6	51.5	58.4	67.8	64.9	56.7	62.0

Table 3. *Cont.*

Method	1-Shot					5-Shot				
	Fold-1	Fold-2	Fold-3	Fold-4	Mean	Fold-1	Fold-2	Fold-3	Fold-4	Mean
RPMMs (ECCV'20)	55.2	66.9	52.6	50.7	56.3	56.3	67.3	54.5	51.0	57.3
PFENet (TPAMI'20)	61.7	69.5	55.4	56.3	60.7	63.1	70.7	55.8	57.9	61.9
ASR (CVPR'21)	53.8	69.6	51.6	52.8	56.9	56.2	70.6	53.9	53.4	58.5
ASGNet (CVPR'21)	58.8	67.9	56.8	53.8	59.3	63.7	70.6	64.2	57.4	63.9
SCL (CVPR'21)	63.0	70.0	56.5	57.7	61.8	64.5	70.9	57.3	58.7	62.9
RePRI (CVPR'21)	59.8	68.3	62.1	48.5	59.7	64.6	71.4	71.7	59.3	66.6
CWT (ICCV'21)	56.3	62.0	59.9	47.2	56.4	61.3	68.5	**68.5**	56.6	63.7
MLC (ICCV'21)	59.2	**71.2**	**65.6**	52.5	62.1	63.5	71.6	71.2	58.1	66.1
HSNet (ICCV'21)	**64.3**	70.7	60.3	**60.5**	**64.0**	**70.3**	**73.2**	67.4	**67.1**	**69.5**
Ours	63.6	70.2	57.1	58.2	62.3	67.7	72.3	59.3	64.1	65.9

Table 4. Comparison of FB-IoU on PASCAL-5^i for 1-shot and 5-shot settings. We used ResNet50 as the backbone.

Methods	1-Shot	5-Shot
PFENet(TPAMI'20)	73.3	73.9
PANet (ICCV'19)	66.5	70.7
CANet (CVPR'19)	66.2	69.6
PGNet (ICCV'19)	69.9	70.5
CRNet (CVPR'20)	66.8	71.5
PPNet (ECCV'20)	69.2	75.8
DAN (ECCV'20)	71.9	72.3
SCL (CVPR'21)	71.9	72.8
ASGNet (CVPR'21)	69.2	74.2
ASR (ICCV'21)	71.3	72.5
Ours	72.5	76.0

Figure 4. Qualitative examples of 5-shot segmentation on the PASCAL-5^i. (**a**) The ground-truth of the query images. (**b**) Results of baseline (PFENet) . (**c**) Results of BGL. (**d**) Results of CPG. (**e**) Results of the combination of BGL and CPG. Best viewed in color and zoomed in.

4.4. Ablation Study

To verify the effectiveness of out proposed methods, we conduct extensive ablation studies with a ResNet-50 backbone on PASCAL-5^i.

4.4.1. The Effectiveness of CPG

To verify the effectiveness of CPG, we conduct several experiments on prototype generation and compare it with other prototype generation algorithms. As a kind of soft cluster algorithm, we first compare our method with Adaptive K-means Algorithm (AK) provided by ASGNet [5], and a traditional algorithm, Expectation-Maximization Algorithm (EM), as shown in Table 5. Compared to the baseline, both AK and EM degenerate the performance of segmentation in a 1-shot setting while our CPG offers 0.6% improvement on the baseline. Compared to SCL [6] which needs to segment both support images and query images, our approach uses less computation cost and inference times (in Table 6) with competitive results on both 1-shot and 5-shot settings. These indicated the superiority of CPG on the few-shot segmentation task.

Table 5. Ablation study on prototype generation in a 1-shot setting on PASCAL-5^i.

Methods	Fold-1	Fold-2	Fold-3	Fold-4	Mean
baseline	61.7	69.5	55.4	56.3	60.8
AK [5]	60.5	68	55	54.2	59.4
EM	56.9	67.7	54.2	53.6	58.1
CPG	62.9	69.6	56.8	56.4	61.4

Table 6. Ablation study on the effectiveness of different components, evaluated on PASCAL-5^i. We report the mIoU and Frames (number of episodes) per second (FPS) for 1-shot and 5-shot. CPG: Complementary Prototypes Generation. BGL: Background Guided Learning.

CPG	BGL	1-Shot	FPS	5-Shot	FPS
-	-	60.7	50	61.9	12.5
√	-	61.4	50	63.6	11.11
-	√	62.1	50	65.1	12.5
√	√	62.3	50	65.9	11.11

4.4.2. The Effectiveness of BGL

To demonstrate the effectiveness of our proposed BGL, we conduct both qualitative and quantitative analysis on BGL. We assume the BGL has two sides of effectiveness on feature representation. The first one is the enhancement of feature representation for the novel classes and the second one is discrimination between the class-specific (foreground) feature and the class-agnostic (background) feature. Following [28], we measure the inter-class variance, intra-class variance, and discriminative function ϕ. Here ϕ is defined as inter-class variance divided by the intra-class variance.

As shown in Figure 5a,b,d, BGL not only enlarges the inter-class variance for novel classes but also increases intra-class variance for novel classes. In other words, BGL does not improve the representation discriminability for novel classes. However, as shown in Figure 5c,e, BGL enlarges the inter-class distance and increases the discriminative function ϕ between the foreground and the background. Therefore, the effectiveness of BGL is in the promotion of discrimination between the foreground and background.

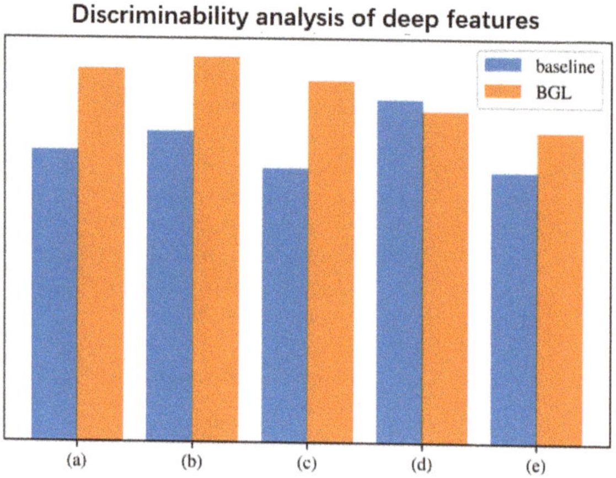

Figure 5. Discriminability analysis. (**a**) intra-class variance on novel classes. (**b**) Inter-class variance on novel classes. (**c**) Inter-class variance on the foreground/background. (**d**) Discriminative function ϕ on the novel class. (**e**) Discriminative function ϕ on the foreground/background.

4.4.3. The Effectiveness of BGL and CPG

To demonstrate the effectiveness of both CPG and BGL, ablation studies are conducted on PASCAL-5^i, as shown in Table 6. Compared with the baseline, using CPG and BGL alone improves the performance by a large margin, 1.7% and 2.6% for mIoU on 5-shot setting, respectively. In addition, we show that using CPG alone could achieve the current SOTA performance provided by SCL [6], and using BGL could surpass thestate-of-the-art performance with a 2.2% mIoU score. Then, combining both CPG and BGL achieves higher performance than the aforementioned one, with 4% improvement in total. In Figure 4, we show that using CPG and BGL alone may generate wrong segmentations on the background, but a combination of them could improve the results. In Figure 6, we show some representative heatmap examples, which further shows how the combination of CPG and BGL helps the model segment precisely and accurately.

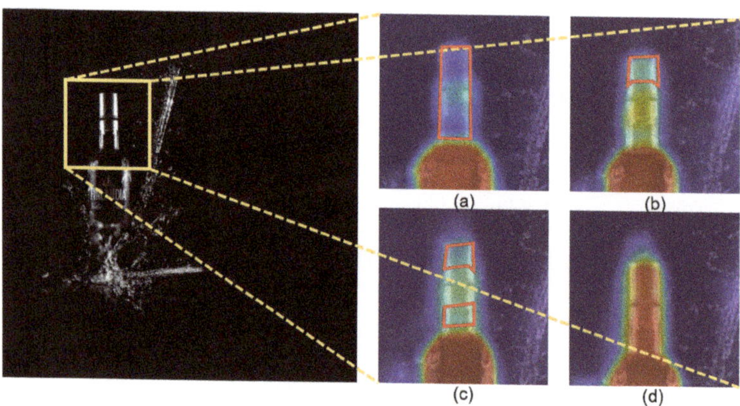

Figure 6. Heatmap examples on PASCAL-5^i in a 5-shot setting. (**a**) Result of baseline. (**b**) Result of CPG . (c) Result of BGL. (**d**) Result of the combination of BGL and CPG.

5. Conclusions

In this paper, we propose a novel few-shot semantic segmentation method named DCNet, which is composed of CPG and BGL. Our approach is able to extract comprehensive support information through our proposed CPG module and generate discriminative feature representation for background pixels by BGL. Extensive experiments demonstrate the effectiveness of our proposed method.

Author Contributions: Main contribution: Q.R.; supervision: J.C. All authors have read and agreed to the published version of the manuscript.

Funding: This work is supported by the Nature Science Foundation of China (No.61972217, No. 62081360152), Natural Science Foundation of Guangdong Province in China (No. 2019B15 15120049, 2020B1111340056).

Institutional Review Board Statement: Ethical review and approval were waived for this study, due to no humans or animals were involved.

Informed Consent Statement: Not applicable.

Data Availability Statement: Not applicable.

References

1. Tian, Z.; Zhao, H.; Shu, M.; Yang, Z.; Li, R.; Jia, J. Prior Guided Feature Enrichment Network for Few-Shot Segmentation. *arXiv* **2020**, arXiv:2008.01449.
2. Liu, Y.; Zhang, X.; Zhang, S.; He, X. Part-Aware Prototype Network for Few-Shot Semantic Segmentation. *arXiv* **2020**, arXiv:2007.06309.
3. Wang, K.; Liew, J.H.; Zou, Y.; Zhou, D.; Feng, J. PANet: Few-Shot Image Semantic Segmentation With Prototype Alignment. In Proceedings of the 2019 IEEE/CVF International Conference on Computer Vision (ICCV), Seoul, Korea, 27 October–2 November 2019; pp. 9196–9205.
4. Sung, F.; Yang, Y.; Zhang, L.; Xiang, T.; Torr, P.H.; Hospedales, T.M. Learning to Compare: Relation Network for Few-Shot Learning. In Proceedings of the 2018 IEEE/CVF Conference on Computer Vision and Pattern Recognition, Salt Lake City, UT, USA, 18–22 June 2018; pp. 1199–1208.
5. Li, G.; Jampani, V.; Sevilla-Lara, L.; Sun, D.; Kim, J.; Kim, J. Adaptive Prototype Learning and Allocation for Few-Shot Segmentation. *arXiv* **2021**, arXiv:2104.01893.
6. Zhang, B.; Xiao, J.; Qin, T. Self-Guided and Cross-Guided Learning for Few-Shot Segmentation. *arXiv* **2021**, arXiv:2103.16129.
7. Zhang, C.; Lin, G.; Liu, F.; Yao, R.; Shen, C. CANet: Class-Agnostic Segmentation Networks With Iterative Refinement and Attentive Few-Shot Learning. In Proceedings of the 2019 IEEE/CVF Conference on Computer Vision and Pattern Recognition (CVPR), Long Beach, CA, USA, 15–20 June 2019; pp. 5212–5221.
8. Yang, B.; Liu, C.; Li, B.; Jiao, J.; Ye, Q. Prototype mixture models for few-shot semantic segmentation. In *Computer Vision—ECCV 2020*, Lecture Notes in Computer Science; Springer: Cham, Switzerland, 2020; Volume 12353, pp. 763–778.
9. Long, J.; Shelhamer, E.; Darrell, T. Fully Convolutional Networks for Semantic Segmentation. In Proceedings of the 2015 IEEE Conference on Computer Vision and Pattern Recognition (CVPR), Boston, MA, USA, 7–12 June 2015; p. 10.
10. Yang, M.; Yu, K.; Zhang, C.; Li, Z.; Yang, K. DenseASPP for Semantic Segmentation in Street Scenes. In Proceedings of the 2018 IEEE/CVF Conference on Computer Vision and Pattern Recognition, Salt Lake City, UT, USA, 18–23 June 2018; pp. 3684–3692.
11. Chen, L.C.; Zhu, Y.; Papandreou, G.; Schroff, F.; Adam, H. Encoder-Decoder with Atrous Separable Convolution for Semantic Image Segmentation. *arXiv* **2018**, arXiv:1802.02611.
12. Zhao, H.; Zhang, Y.; Liu, S.; Shi, J.; Loy, C.C.; Lin, D.; Jia, J. PSANet: Point-Wise Spatial Attention Network for Scene Parsing. In Proceedings of the Proceedings of the European Conference on Computer Vision (ECCV), Munich, Germany, 8–14 September 2018; pp. 267–283.
13. Zhang, H.; Dana, K.; Shi, J.; Zhang, Z.; Wang, X.; Tyagi, A.; Agrawal, A. Context Encoding for Semantic Segmentation. In Proceedings of the Proceedings of the IEEE Conference on Computer Vision and Pattern Recognition, Salt Lake City, UT, USA, 18–23 June 2018; pp. 7151–7160.
14. Zhang, H.; Zhang, H.; Wang, C.; Xie, J. Co-Occurrent Features in Semantic Segmentation. In Proceedings of the IEEE/CVF Conference on Computer Vision and Pattern Recognition, Long Beach, CA, USA, 15–20 June 2019; pp. 548–557.
15. Yuan, Y.; Huang, L.; Guo, J.; Zhang, C.; Chen, X.; Wang, J. OCNet: Object Context Network for Scene Parsing. *arXiv* **2021**, arXiv:1809.00916.
16. Fu, J.; Liu, J.; Tian, H.; Li, Y.; Bao, Y.; Fang, Z.; Lu, H. Dual Attention Network for Scene Segmentation. *arXiv* **2019**, arXiv:1809.02983.
17. Huang, Z.; Wang, X.; Huang, L.; Huang, C.; Wei, Y.; Liu, W. CCNet: Criss-Cross Attention for Semantic Segmentation. In Proceedings of the IEEE/CVF International Conference on Computer Vision, Seoul, Korea, 27–28 October 2019; pp. 603–612.
18. Chen, W.Y.; Liu, Y.C.; Kira, Z.; Wang, Y.C.F.; Huang, J.B. A Closer Look at Few-Shot Classification. *arXiv* **2020**, arXiv:1904.04232.
19. Vinyals, O.; Blundell, C.; Lillicrap, T.; Kavukcuoglu, K.; Wierstra, D. Matching Networks for One Shot Learning. *arXiv* **2017**, arXiv:1606.04080.
20. Snell, J.; Swersky, K.; Zemel, R.S. Prototypical Networks for Few-Shot Learning. *arXiv* **2017**, arXiv:1703.05175.

21. Finn, C.; Abbeel, P.; Levine, S. Model-Agnostic Meta-Learning for Fast Adaptation of Deep Networks. *arXiv* **2017**, arXiv:1703.03400.
22. Shaban, A.; Bansal, S.; Liu, Z.; Essa, I.; Boots, B. One-Shot Learning for Semantic Segmentation. *arXiv* **2017**, arXiv:1709.03410.
23. Liu, W.; Zhang, C.; Lin, G.; Liu, F. CRNet: Cross-Reference Networks for Few-Shot Segmentation. In Proceedings of the 2020 IEEE/CVF Conference on Computer Vision and Pattern Recognition (CVPR), Seattle, WA, USA, 13–19 June 2020; pp. 4164–4172. [CrossRef]
24. Zhang, X.; Wei, Y.; Yang, Y.; Huang, T. SG-One: Similarity Guidance Network for One-Shot Semantic Segmentation. *arXiv* **2020**, arXiv:1810.09091.
25. Yang, L.; Zhuo, W.; Qi, L.; Shi, Y.; Gao, Y. Mining Latent Classes for Few-Shot Segmentation. *arXiv* **2021**, arXiv:2103.15402.
26. Chen, J.; Gao, B.B.; Lu, Z.; Xue, J.H.; Wang, C.; Liao, Q. SCNet: Enhancing Few-Shot Semantic Segmentation by Self-Contrastive Background Prototypes. *arXiv* **2021**, arXiv:2104.09216.
27. Nguyen, K.; Todorovic, S. Feature Weighting and Boosting for Few-Shot Segmentation. In Proceedings of the 2019 IEEE/CVF International Conference on Computer Vision (ICCV), Seoul, Korea, 27 October–2 November 2019; pp. 622–631. [CrossRef]
28. Liu, B.; Cao, Y.; Lin, Y.; Li, Q.; Zhang, Z.; Long, M.; Hu, H. Negative Margin Matters: Understanding Margin in Few-Shot Classification. In *Computer Vision—ECCV 2020*; Vedaldi, A., Bischof, H., Brox, T., Frahm, J.M., Eds.; Springer International Publishing: Cham, Switzerland, 2020; Volume 12349, pp. 438–455. [CrossRef]

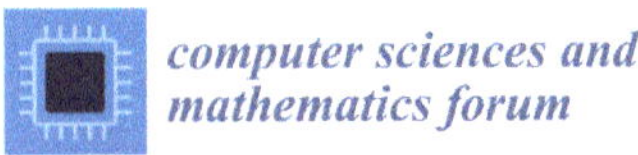
computer sciences and mathematics forum

Proceeding Paper

Extracting Salient Facts from Company Reviews with Scarce Labels †

Jinfeng Li [1,*], Nikita Bhutani [1], Alexander Whedon [2,‡], Chieh-Yang Huang [3,‡], Estevam Hruschka [1] and Yoshihiko Suhara [1]

1 Megagon Labs, Mountain View, CA 94041, USA; nikita@megagon.ai (N.B.); estevam@megagon.ai (E.H.); yoshi@megagon.ai (Y.S.)
2 Stitch Fix, San Francisco, CA 94104, USA; alexander.whedon@gmail.com
3 College of Information Sciences and Technology, Pennsylvania State University, State College, PA 16801, USA; chiehyang@psu.edu
* Correspondence: jinfeng@megagon.ai
† Presented at the AAAI Workshop on Artificial Intelligence with Biased or Scarce Data (AIBSD), Online, 28 February 2022.
‡ A.W.: Work done while at Megagon Labs. C.-Y.H.: Work done during internship at Megagon Labs.

Abstract: In this paper, we propose the task of extracting salient facts from online company reviews. Salient facts present unique and distinctive information about a company, which helps the user in deciding whether to apply to the company. We formulate the salient fact extraction task as a text classification problem, and leverage pretrained language models to tackle the problem. However, the scarcity of salient facts in company reviews causes a serious label imbalance issue, which hinders taking full advantage of pretrained language models. To address the issue, we developed two data enrichment methods: first, representation enrichment, which highlights uncommon tokens by appending special tokens, and second, label propagation, which interactively creates pseudopositive examples from unlabeled data. Experimental results on an online company review corpus show that our approach improves the performance of pretrained language models by up to an F1 score of 0.24. We also confirm that our approach competitively performs well against the state-of-the-art data augmentation method on the SemEval 2019 benchmark even when trained with only 20% of training data.

Keywords: review mining; natural language processing; information extraction; pretrained models; scarce labels

Citation: Li, J.; Bhutani, N.; Whedon, A.; Huang, C.-Y.; Hruschka, E.; Suhara, Y. Extracting Salient Facts from Company Reviews with Scarce Labels. *CSFM* **2022**, *3*, 9. https://doi.org/10.3390/cmsf2022003009

Academic Editors: Kuan-Chuan Peng and Ziyan Wu

Published: 29 April 2022

Publisher's Note: MDPI stays neutral with regard to jurisdictional claims in published maps and institutional affiliations.

1. Introduction

Online reviews are an essential source of information. More than 80% of people read online reviews before reaching decisions [1]. This trend also applies to job seekers. Before applying to open positions, job seekers often read online employee reviews about hiring experience and work environment on Indeed, LinkedIn, and other channels. However, the overabundance of reviews can render them cumbersome to read. For example, there are 63,400 reviews about Amazon on Indeed. Furthermore, job seekers must skim through several subjective comments in the reviews to find concrete information about a company of interest.

Alternatively, job seekers can find such concrete information (e.g., Table 1) in expert articles about companies on websites such as Business Insider [2,3] and FutureFuel [4]. However, such expert articles are typically written only for very popular companies and do not cover the global majority of companies. Online company reviews, on the other hand, are available for a vast number of companies, as (former) company employees submit reviews about a company to review platforms such as Glassdoor. Therefore, we aim to automatically extract unique and distinctive information from online reviews.

We refer to informative descriptions in online reviews as salient facts. In order to derive a formal definition of salient facts, we conducted an inhouse study where we asked three editors to inspect 43,000 reviews about Google, Amazon, Facebook, and Apple. The editors discussed salient and nonsalient sentences in the reviews, and concluded that a salient fact mentions an uncommon attribute about a company and/or describes some quantitative information of an attribute. Attributes of a company include employee perks, onsite services and amenities, the company culture, and the work environment. We further validated our definition by looking into expert articles, and confirmed that the articles were extensively composed of the same properties. For example, 4 of the 8 benefits mentioned in an article [2] about Google used less-known attributes such as food variety, fitness facilities, and pet policy. The other 4 of 8 benefits used numeric values, such as 50% retirement pension match.

Table 1. Sample sentences from an online review and expert article about Google.

Online	Good work place, best pay, awesome company.
Expert	In the event of your death, Google pays your family 50% of your salary each year.

In this paper, we propose the novel task of salient fact extraction and formulate it as a text classification problem. With this formulation, we could automate filtering company reviews that contain salient information about the company. Pretrained models [5–7] are a natural choice for such tasks [8,9] since they generalize better when the training data for the task are extremely small. We, therefore, adopted BERT [5] for our extraction task. However, generating even a small amount of task-specific balanced training data is challenging for salient fact extraction due to the scarcity of salient sentences in the reviews. Naively labeling more sentences to address the scarcity can be prohibitively expensive. As such, even pretrained models that perform robustly in few-shot learning cannot achieve good enough performance when used directly for this task.

In this work, we propose two data enrichment methods, representation enrichment and label propagation, to address the scarcity of salient facts in training data. Our representation enrichment method is based on the assumption that salient sentences tend to mention uncommon attributes and numerical values. We can, therefore, enrich training data using automatically identified uncommon attributes and numeric descriptions from review corpora. Specifically, we append special tags to sentences that mention uncommon attributes and numerical values to provide additional signals to the model. Our label propagation method is based on the idea that we can use a small set of seed salient sentences to fetch similar sentences from unlabeled reviews that are likely to be salient. This can help in improving the representation of salient sentences in the training data. Our methods are applicable to a wide variety of pretrained models [5–7].

We conducted extensive experiments to benchmark the extraction performance and demonstrate the effectiveness of our proposed methods. Our methods could improve the F1 scores of pretrained models by up to 0.24 on salient fact extraction, which is 2.2 times higher than the original F1 scores. This is because our models could identify more uncommon attributes and more quantitative descriptions than directly using pretrained language models can.Our models could also better distinguish between expert- and employee-written reviews.

To summarize, our contributions are the following: (1) We practice a new review mining taskcalled salient fact extraction using pretrained language models and data augmentation in an end-to-end manner. The task faced an extremely low ratio (i.e., <10%) of salient facts in raw reviews. (2) The best-performing methods still require massive labels in the tens of thousands, because trained models and augmented examples tend to be biased towards majority examples. To alleviate this problem, we leveraged a series of improvements to ensure that the model training and data augmentation worked effectively for the desired minority examples. (3) An extension of our method demonstrates that it generalizes well and could reduce the labeling cost for new domain adaption (e.g., transferring to a product domain achieves an improved label ratio from 5% to 43%) of the same task and for similar tasks that deal with minority review comment extraction (e.g., suggestion mining requires a reduced amount of labels by 75% to hit the performance of UDA semisupervised learning [10]). To facilitate future research, we publicized our implementations and experimental scripts (https://github.com/megagonlabs/factmine, accessed on 22 August 2020). We did not release the company dataset due to copyright issues. However, we aim to release datasets of similar tasks to benchmark the performance of different methods. We also released a command-line programming interface that renders our results readily reproducible.

2. Characterization of Salient Facts

The cornerstone towards automatic extraction is to understand what renders a review (or sentence in a review) salient. To this end, we first inspected raw online reviews to derive a definition of salient facts. We then analyzed expert articles to ensure that the derived definition is valid (Figure 1).

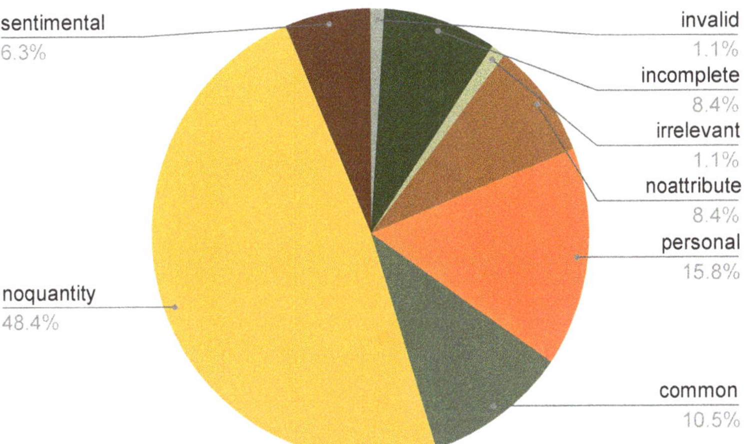

Figure 1. Constitution of false instances.

2.1. Review Corpus Annotation and Analysis

We produced inhouse annotation to understand what review sentences are deemed salient facts for human readers.We collected 43,000 company reviews about Google, Amazon, Facebook, and Apple. We split each review into sentences using NLTK [11]. Then, we inspected all the sentences and selected salient sentences according to our understanding of the corresponding companies. Table 2 shows example sentences that were labeled salient.

Sentences labeled salient described more uncommon attributes than nonsalient sentences did. Uncommon attributes include real-world objects and services such as cafes, kitchens, dog parks. They are typically not provided by all companies and can help job seekers differentiate between companies. Furthermore, salient sentences use quantitative descriptions (e.g., 25+ and 100 ft in Table 2). Quantities often represent objective information and vary across companies, even for the same attribute, thereby helping job seekers in differentiating between companies.

These properties are not exhibited by nonsalient sentences. As shown in Table 3, most nonsalient sentences mention solely common attributes (e.g., place, salary and people), disclose purely personal sentiments (e.g., awesome, great, cool), or are noisy (e.g., invalid or incomplete texts). Different kinds of nonsalient sentences and their ratios are shown in Figure 1.

Table 2. Sample salient facts extracted from online reviews.

Example 1. Google also has 25+ cafes and microkitchens every 100 ft. (Google)
Example 2. Dogs allowed in all the buildings I've been to (including some dog parks in the buildings!) (Amazon)

Table 3. Example non-salient sentences and reasons.

Reason	Example
noquantity	awesome place to work, great salary, smart people
personal	I couldn't imagine a better large corporate culture that still tries to be agile
common	Salary, perks, and benefits
noattribute	ok ok ok ok ook
incomplete	five single words for this
sentimental	great, happy, cool, friendly, doable, beautiful, awesome, nice, good, big
invalid	good fv gt tr tr yt y
irrelevant	Best friendly free cab cool no target

2.2. Expert Article Analysis

We analyze expert-written reviews to investigate if they exhibited characteristics of salient facts i.e., describe an uncommon attribute and/or use quantitative descriptions. First, we compare frequencies of a set of attribute words across expert sentences and review sentences. The used expert sentences attributed words that were infrequent in the review sentences. For example, frequencies of *death, family* (commonly mentioned in expert reviews for Google) in review sentences were 0.01% and 0.15%, respectively. In contrast, frequencies of *place, pay* (commonly mentioned in review sentences for Google) were 3.44% and 1.28%, respectively. This observation supports our definition.

Next, we inspected if the expert sentences used more quantitative descriptions than randomly selected review sentences. For example, 4 of the 7 expert sentences describing most benefits of Google used quantitative descriptions such as 10 years, USD 1000 per month, 18–22 weeks, and 50% match. On the other hand, none of the 7 sentences randomly sampled from reviews mentioned any quantities. In fact, most of them used subjective descriptions such as nice, interesting, and great. This observation supports our characterization of salient facts.

3. Methodology

Owing to the recent success of pretrained models in information extraction tasks, we adopted these models for salient fact extraction. We first describe how we modelled salient fact extraction as a sentence classification task over pretrained models. We describe technical challenges unique to this task. We then describe two methods, *representation enrichment* and *label propagation*, to address these challenges.

3.1. Pretrained Model and Fine Tuning

The goal of a supervised-learning model for salient fact extraction tasks is to predict the correct label for an unseen review sentence: 1 if the sentence is salient, and 0 otherwise. The model is trained using a set of labeled text instances $(t, l)_i$, where t is a sentence and l is a binary label. By seeing a number of training instances, the model learns to discriminate between positive and negative instances. However, supervised learning is sensitive to the coverage of salient sentences in the review corpus. It can yield suboptimal models when faced with imbalanced datasets.

Pretrained models, on the other hand, tend to be more robust to such imbalances and generalize better. These models project a text instance t into a high-dimensional vector (e.g., 768 in BERT), such that text instances sharing similar words or synonyms have similar vectors. Since predictions are based on dense-vector representations, they can predict the same label for semantically equivalent instances (e.g., cafe and coffee) without having seen them explicitly during training. As a result, pretrained models require far fewer salient sentences than supervised models trained from scratch do.

Despite their better generalizability, pretrained models struggle to make correct predictions for sentences with unseen attributes or quantities if their synonyms didn't appear in the training set. As a result, a training set should contain as many infrequent attributes and quantitative descriptions as possible for optimal performance of pretrained models. However, due to the inherent scarcity of infrequent attributes and quantitative descriptions, the models can only see a limited amount of salient facts (and thus infrequent attributes and quantities) during training. We propose representation enrichment and label propagation methods to address these challenges. We next describe these methods in more detail.

3.2. Representation Enrichment

In our empirical experiment, we observed that only 0.55% of labeled sentences were considered to be positive (i.e., salient facts.). Given such an extremely small number of positive examples, there is a chance that the learning algorithm cannot generalize the model using the training set as it may not cover sufficient patterns of salient facts. As a result, a trained model may not be able to recognize salient facts with different linguistic patterns than those of the training instances. In this paper, we considered that we could alleviate the issue by incorporating prior knowledge about the task. Salient facts contain relatively uncommon attributes and/or quantitative descriptions, so we aimed to implement those functions into the model.

Since models may meet unseen salient facts during prediction, we developed a representation enrichment method to help the models in recognizing their attributes and quantities for prediction. The method appends a special tag to text instances if they contain tokens related to uncommon attributes or quantitative descriptions. The model can learn that a text instance containing the special tag tends to be a salient fact. During prediction, even if a model does not recognize unseen tokens in a salient fact instance, the model can recognize the special tag and make accurate prediction.

The expansion process begins by selecting a set of salient tokens. The salient tokens are those common words that appear in the review corpus to describe uncommon attributes or quantities. The expansion process comprises two steps. The first step identifies a list of salient tokens as part of the inputs to Algorithm 1. The second step takes the list and a special tag token (e.g., "salient" for uncommon attribute token list) to run Algorithm 1. The algorithm iterates all text instances. If a text instance contains any token of the list, the algorithm appends a special tag to it. All instances that contain salient tokens share the same tag. After the two steps, both groups of tagged and untagged text instances are fitted to train the extraction model. After the model is trained, it is used to produce predictions for salient facts.

Uncommon attribute token list: we used a two-step method to discover tokens that are used in the corpus to describe uncommon attributes. First, we identified nouns, since attribute tokens are mostly nouns. We used NLTK to extract noun words. Second, we ranked the nouns by their IDF scores. The IDF of a noun w is calculated as $\log(T/|\{d|d \in D \wedge w \in d\}|)$, where T is the total number of sentences and $|\{d|d \in D \wedge w \in d\}|$ is the number of sentences that contain the noun token w. Nouns that appeared the least frequently (i.e., top 1000 words based on the IDF scores) in the review corpus were considered to be uncommon attribute tokens. We next inspect the top list to label tokens that are used to describe uncommon attributes. The purpose was to exclude nonattribute words. By applying this two-step method, we successfully constructed a list of uncommon attribute tokens.

Quantitative description token list: we curated a list of tokens that are used to describe quantities. The list contains three types of tokens: digit, numeric, and unit. Digit tokens include all integer numbers from 0 to 9 and any integers composed of the 10 integers. Numeric (https://helpingwithmath.com/cha0301-numbers-words01/, accessed on 22 August 2020) are word descriptions of numbers, and representatives are hundreds, thousands, and millions. Unit (https://usma.org/detailed-list-of-metric-system-units-symbols-and-prefixes, accessed on 22 August 2020) consist of commonly used measurements that often appear in quantitative descriptions, and some examples include hour and percentage. Digit, numeric, and unit form a comprehensive coverage of word tokens that people commonly use in quantitative descriptions. We last inspected the set of tokens and curated a final list of tokens for quantitative descriptions.

Algorithm 1 Representation enrichment.

Input: Text instance t with tokens $t_1, t_2, \ldots, t_k$, list l of salient tokens, and special token s
Output: A new text instance t_{new}

$t_{new} \leftarrow t$
for $i \leftarrow 1$ to k **do**
  if $t_i \in l$ **then**
    $t_{new} \leftarrow t_{new} + s$
    return t_{new}
  end if
end for
return t_{new}

3.3. Label Propagation

Due to the extremely sparse positive examples for salient facts, the training procedure may fail to generalize the model. To alleviate the issue, we augmented training data by searching similar instances.

Candidate Selection: we show the label propagation process in Algorithm 2. The process takes salient fact instance t from existing training data as input. Then, it searches the m-most similar instances from unlabeled text instances (denoted as $u_1, u_2, \ldots, u_n$.) As the similarity function, we used the Jaccard score as defined in Equation (1), where V_t and V_u denote the distinct vocabulary sets of t and u respectively. The score is 1 if two texts share exactly same vocabulary sets, and 0 if they do not share any common tokens.

$$J(t,u) = |V_t \cap V_u| / |V_t \cup V_u| \tag{1}$$

To obtain vocabulary sets, we used the BERT WordPiece tokenizer to split the text into tokens by matching character sequence with a predefined vocabulary of about 30,000 [5]. Since an unlabeled corpus contains abundant text instances, Algorithm 2 can help in retrieving the instances that are the most similar to salient facts to expand our training set.

Algorithm 2 Label propagation: candidate selection.

Input: Salient instances set T, unlabeled instances set U, similarity function $sim(t,u)$, candidate size m
Output: Candidate instances set C of size m

Candidate set $C \leftarrow [\,]$
for t in T **do**
  for $i \leftarrow 1$ to n **do**
    $s \leftarrow sim(t, u_i)$
    $C \leftarrow C + (u_i, s)$
  end for
end for
$C = deduplicate(C)$
Sort C by score
return $C[1:m]$

Reranking: Jaccard score favors frequent word tokens such as stopwords. Therefore, a negative instance can be ranked high and returned as a candidate if it contains a lot of stopwords. To solve this issue, we introduced a reranking operator that sorts all candidates by their relative affinity to positive and negative examples in the training set, as shown in Algorithm 3. For every candidate c, we calculated two scores, i.e., textual affinity ta and semantic affinity sa, which were used to measure the overall distances to a group of examples G.

$$avg\_dist(c, G) = 1/|G| * \sum_{e \in G}(1 - J(c, e)) \tag{2}$$

$$ta(c, X, Y) = \frac{avg\_dist(c, \{x_i | x_i \in X, y_i = 0\})}{avg\_dist(c, \{x_i | x_i \in X, y_i = 1\})} \tag{3}$$

Textual affinity ta was defined as Equation (3) to measure the relative affinity of a candidate c to the positive- and negative-example groups of the training set. Affinity is measured by counter average distance (see Equation (2)). Greater textual affinity is better, which means that c has smaller distance to the positive group and larger distance to the negative group. Intuitively, textual affinity favors candidate c that shares many common tokens with positive examples, while such tokens are not common (e.g., stopwords) in negative examples.

$$sa(c, X, Y) = discriminator(X, Y).estimate(c) \tag{4}$$

Textual affinity cannot recognize semantically connected words (e.g., million and billion). Therefore, we introduced semantic affinity sa as defined in Equation (4). Semantic affinity requires a discriminator that uses word embeddings as input representation. In other words, a discriminator can recognize semantically connected words through similar word vectors. Next, we trained the discriminator using the training set, so that the discriminator learned to predict whether an input sentence is a positive example according to its word vectors. The trained discriminator is used to estimate the probability of candidate c belonging to the positive group. In our experiments, we used BERT as the discriminator and took the product of textual affinity ta and semantic affinity sa to yield the best F1 scores.

Lastly, we sorted all candidates in descending order by their overall affinity score (i.e., textual affinity $\times$ semantic affinity). We returned the top pk as positive examples, and tail nk as negative examples, where pk and nk are user-defined parameters. In our experiments, label propagation performed reasonably well if $\frac{pk}{pk+nk}$ equalled to the label ratio, and $pk + nk$ equalled to the training size but was smaller than half the size of unlabeled examples.

Algorithm 3 Label propagation: reranking.

Input: Candidate collection C, training set X, Y, number of pseudopositive examples pk, and negative examples nk

Output: pk positive and nk negative pseudoexamples

```
Reranking set R ← [ ]
for c in C do
    ta = textual_affinity(c, X, Y)
    estimator = BERT (X, Y)
    sa = semantic_affinity(c, estimator)
    R ← R + (c, ta * sa)
end for
R.sorted(key = lambda(c, s) : −s)
return head pk and tail nk of R as positive and negative pseudoexamples
```

3.4. Additional Training Techniques

Fine tuning pretrained language models is limited in batch size due to GPU memory capacity. For example, the maximal batch size that BERT base model can process on a 16 GB GPU is around 64. Given the extremely low label ratio (e.g., 5%), it is possible that a batch may not contain any positive examples. Consequently, the trained model may exhibit significant biases against positive examples. To alleviate this problem, we leveraged two fine-tuning techniques, namely, thresholding and choosing the best snapshot (described below), to enable the trained model to weigh more on the positive examples.

Thresholding: pretrained models such as BERT adopt argmax to predict the label of an example. First, the pretrained model outputs two probability scores for the same example, indicating the likelihood of this example belonging to the negative or positive class. Next, argmax selects the class of a larger score as the final prediction. Experiments showed that the average positive probability was much smaller than negative probability; thus, we replaced argmax with thresholding that only concerned the positive prediction score. Thresholding sorts all examples by positive prediction scores and varies a threshold from the highest to the lowest score. We tried 100 different thresholds at equal intervals between highest and lowest, and chose the threshold that led to the largest F1 on the training set.

Choose best snapshot: due to severe label imbalance, a model could achieve the best performance during its training snapshots. A potential reason is that the model met the highest-quality positive and negative examples at the snapshots. Therefore, we set a fixed number of snapshots and inspected the model during each snapshot. We compared the model performance between two consecutive snapshots and checkpointed the model if better performance was observed.

4. Experiments

In this section, we first examine the extraction performance of pretrained models BERT, ALBERT, and RoBERTa. We then show the effectiveness of our proposed data enrichment methods by conducting an ablation study with the pretrained models.

Datasets: we obtained company reviews from an online company review platform for job seekers. We use the reviews of two companies (Google and Amazon) for evaluation. We chose these companies because their expert articles were also available for comparison. We first split the reviews into sentences using the NLTK sentence tokenizer [11]. For Google, we used all 13,101 sentences from the reviews. For Amazon, we randomly sampled 10,000 sentences. We then asked four editors to finish labeling these sentences (1 or 0) on the basis of their salience. We randomly sampled 100 sentences (50 positive and 50 negative) and asked two editors to label them. There was Cohen's kappa agreement of 0.9 between the editors. This agreement is higher than the agreement scores reported in previous studies related to our work e.g., 0.81 from a SEMEVAL-2019 Competition task 9 [8] and 0.59 from TipRank [12]).

Hyperparameters: we split the labeled dataset into training and test sets at a ratio of 4:1. For training the pretrained models, we set the number of epochs to 5, max sequence length to 128, and batch size to 32. We used the F1 score of the positive class (i.e., salient) to measure the performance of a model. Since a model may achieve the best F1 score in the middle of training, we inspected a model 15 times during training and reported the best F1 score of the 15 snapshots.

4.1. *Effectiveness of Pretrained Models*

We first compare the performance of pretrained models and other supervised learning algorithms, namely, logistic regression (LR), support vector machine (SVM), convolutional neural network (CNN), and recurrent neural network with long short-term memory (LSTM). We used the same configuration to train and evaluate all models. Unsurprisingly, all pretrained models consistently outperformed other models on the two datasets (as shown in Table 4). BERT achieved the highest F1 scores with absolute F1 gain as high as 0.16 and 0.14 on Google and Amazon, respectively. These results indicate that the pretrained models are suited for the salient fact extraction task.

Table 4. F1 scores of BERT, ALBERT (ALB.), RoBERTa (ROB.), LR, SVM, CNN, and LSTM on Google and Amazon datasets. The best score for each dataset is in bold.

Dataset	BERT	ALB.	ROB.	LR	SVM	CNN	LSTM
Google	**0.33**	0.30	0.19	0.13	0.17	0.17	0.17
Amazon	**0.27**	0.13	0.20	0.13	0.12	0.03	0.07

4.2. Effectiveness of Representation Enrichment

To investigate the effectiveness of representation enrichment, we curated two lists, one for uncommon attribute descriptions and one for quantitative descriptions. We separately applied the two lists for each pretrained model, and report their F1 scores in Table 5. We also computed the F1 scores before and after representation enrichment.

Table 5. F1 score of BERT, ALBERT (ALB.), RoBERTa (ROB.) when using representation enrichment. F1 improvements compared with direct use of pretrained models (see Table 4) marked in orange. Best scores marked in bold.

Expansion	Dataset	BERT	ALB.	ROB.
Uncommon	Google	0.38 (+0.05)	**0.43** (+0.13)	0.19 (+0.00)
Uncommon	Amazon	0.29 (+0.02)	0.28 (+0.15)	**0.35** (+0.15)
Quantitative	Google	0.38 (+0.05)	**0.40** (+0.10)	0.32 (+0.13)
Quantitative	Amazon	0.27 (+0.00)	0.2 (+0.7)	**0.44** (+0.24)

We first evaluated the effect of representation enrichment using uncommon attribute token list (Uncommon). As shown in Table 5, Uncommon could improve the F1 score of BERT, which appeared to be the best model, as shown in Table 4, from 0.33 to 0.38 on Google and from 0.27 to 0.29 on Amazon, so improvement was 0.05 and 0.02, respectively. More importantly, Uncommon also improved the F1 scores of models ALBERT and RoBERTa on both Google and Amazon. ALBERT achieved the greatest F1 improvement (0.13 on Google and 0.15 on Amazon) and outperformed BERT. RoBERTa achieved 0.15 F1 improvement and outperformed BERT on Amazon. Results indicate that representation enrichment with an uncommon attribute token list is generic and can improve the extraction performance of various pretrained models.

We next evaluated the effect of representation enrichment using quantitative description token list (Quantitative). As shown in Table 5, Quantitative consistently improved F1 scores for all models. In particular, ALBERT achieved F1 improvement of 0.10 on Google, while RoBERTa an F1 improvement of 0.24 on Amazon. The final F1 score of RoBERTa was 0.44 on Amazon, and the score was record-high in Amazon extraction performance. Results further verified that representation enrichment, in particular the quantitative description token list, is a general method that works with various pretrained models.

4.3. Effectiveness of Label Propagation

Label propagation boosts the number of training samples by retrieving similar texts from unlabeled corpora. To evaluate the effect of label propagation, we retrieved three of the most similar texts for each salient fact and use them as positive examples for training. Since Google and Amazon had 62 and 66 salient facts, we retrieved 186 and 198 sentences, respectively. We report the F1 scores of BERT, ALBERT, and RoBERTa in Table 6. We also calculated the F1 improvements before and after the label propagation.

Table 6. F1 score of BERT, ALBERT (ALB.), RoBERTa (ROB.) when using label propagation. F1 improvement compared with direct use of pretrained models (see Table 4) are marked in orange. Best scores are marked in bold.

Dataset	BERT	ALB.	ROB.
Google	**0.48** (+0.15)	0.37 (+0.07)	0.36 (+0.17)
Amazon	0.28 (+0.01)	0.22 (+0.09)	**0.29** (+0.07)

Pretrained models achieved better F1 scores with label propagation. F1 improvement ranged from 0.07 to 0.17 on Google, and 0.01 to 0.09 on Amazon. RoBERTa showed the largest improvement of 0.17 on Google, where its F1 score rose up to 0.36 from 0.19, which did not leverage label propagation (see Table 4). On Google, BERT achieved0.15 F1 improvement and a record-high F1 score of 0.48. Results suggest that label propagation can boost the performance of various pretrained models.

5. Extension

In this section, we extend our method to a new domain and similar tasks that deal with imbalanced datasets to verify whether our task and method had much generality.

5.1. New Domain

We defined the concept of salient fact from analyzing company reviews. We then attempted to transfer the concept to a new domain, i.e., product reviews. First, we directly deployed a trained company model on product review sentences to predict their probability of saliency. Next, we sorted all sentences by saliency score in descending order, and present the top 100 to 4 human annotators. We asked annotators to give label every sentence with positive or negative indicating salient or nonsalient, respectively. We also asked annotators to label randomly sampled 100 sentences for comparison.

We report the averaged ratio of positive examples for four headset products, i.e., plantronic, jawbone, Motorola, and Samsung, in Table 7. According to the results, transferring consistently increased the label ratio by a large margin for all four products. The margin varied from 3× to 7×. Results suggest that the definition of salient facts is general enough to be applied to the product domain. For quick demonstration, we release all sentence samples in our public codebase.

Table 7. Ratio of sentences that human annotators feel salient before and after transferring trained company model to product reviews.

	Plantronic	Jawbone	Motorola	Samsung
random	0.05	0.08	0.08	0.06
transfer	0.40	0.37	0.38	0.39

5.2. Similar Public Task

We extended the label propagation algorithm to similar tasks since the algorithm was designed to be general. We conducted experiments to compare our method with the state-of-the-art baselines on public tasks that regard minority comment extraction. We obtained four public datasets that contained binary labels for training extraction models. SUGG [8] comes from SEMEVAL 2019 task 9; positive example means that it contains customer suggestions for software improvement. HOTEL [13] was derived from the Hotel domain with, positive example indicating that it carries customer-to-customer suggestions for accommodation. SENT [14] contains sentence-level examples, and a positive label means the sentence contains tips for PHP API design. PARA [14] comes from the same source of SENT, but contain paragraph-level examples. The ratio of positive examples for SUGG, HOTEL, SENT, and PARA was 26%, 5%, 10%, 17%, respectively. All four datasets contained a training set and a test set at 4:1 ratio.

We adopted UDA [10] as a strong baseline method. UDA uses BERT as base model and augments every example in the training set using back translation from English to French then back to English. The example and its back translation are fed into model training to minimize KL divergence, so that the two examples are projected to close vector representations. We ran UDA and BERT on the full training set, and our method on only 2000 training examples. Our F1 scores and those of BERT and UDA are shown Table 8. The average F1 of BERT, UDA, and ours was 0.6687, 0.6980, and 0.6961, respectively. BERT performed the worst because it does not use any data augmentation, so it suffers the most from label imbalance. UDA and ours performed similarly across all the datasets, yet UDA used full training examples, but ours used only 23.52%, 33.33%, 21.97%, and 38.46% of the examples on SUGG, HOTEL, SENT, and PARA, respectively. UDA favors mild data augmentation due to the usage of KL divergence and back translation mostly change one or two word tokens in an example. However, the mild design choice was too conservative to efficiently augment minority examples in imbalanced datasets (thus requiring a higher volume of augmented data). Therefore, a more aggressive design choice such as ours, which can return new sentences as augmented examples, is needed for the widespread existence of imbalanced datasets.

Table 8. F1 of four public tasks for minority comment extraction. All baselines use full training examples. Our method used 2000, yet could match the performance of baselines.

	SUGG (8.5k)	HOTEL (6k)	SENT (9.1k)	PARA (5.2k)
BERT (full)	0.8571	0.6467	0.5413	0.6297
UDA (full)	0.8695	0.7290	0.5614	0.6322
Ours (2k)	0.8673	0.7244	0.5416	0.6514

5.3. Statistical Significance

We conducted experiments to evaluate the statistical significance or randomness of our results. Specifically, we set different random seeds to run BERT, UDA, and our method on SUGG, HOTEL, SENT, and PARA. The number of training examples for SUGG, HOTEL, SENT, and PARA was 8500, 6000, 9100, and 5200, respectively. For every dataset, we fed full training examples to BERT and UDA, but only 2000 to our method. We repeated the same experiment three times and reported F1 scores. Statistical analysis was performed using GraphPad Prism 7, and statistical significance was determined using one-way ANOVA followed by Tukey's multiple-comparison test. We calculated the mean, SD, and p value with Student's t test. Significance: not significant (n.s.) $p > 0.5$, * $p < 0.05$, ** $p < 0.01$, *** $p < 0.001$.

Comparison results of BERT, UDA, and ours (2000) on SUGG, HOTEL, SENT, and PARA shown in Figure 2. When comparing BERT with ours (2000), BERT showed no significant difference on SUGG and SENT, and worse performance on HOTEL and PARA. Results suggest that ours (2000) could outperform BERT even with fewer training examples. When comparing UDA and ours (2000), the two methods showed no significant difference on SENT and PARA. On HOTEL, UDA was better, but on SUGG it showed worse performance. Results suggest that ours (2000) could achieve equally good performance as that of UDA with much fewer training examples.

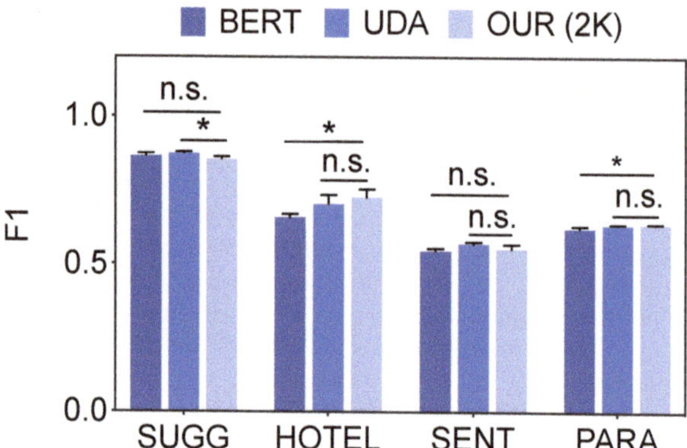

Figure 2. Comparison between BERT and UDA, with our method. BERT and UDA are trained with full training examples, and our method was trained with only 2000 examples. Training datasets were SUGG, HOTEL, SENT, and PARA. Data are presented as *mean* ± *SD*. Significance: not significant (n.s.) $p > 0.5$, * $p < 0.05$.

6. Related Work

Informative reviews: extracting informative reviews drives broad applications in web mining, while the definition of informativeness varies across application domains. TipRank [12] extracts short and practical sentences from TripAdvisor reviews to prepare travellers for upcoming trips. AR-Miner [15] and DeepTip [14] highlight useful comments in software reviews to notify developers of potential artifact issues. AMPERE [16] extracts argumentative sentences from paper reviews to help authors improve their manuscripts. In addition to the above research, there are many works targeting different domains such as products [17–19], restaurants [20–22], and hotels [13,23,24]. These works align with discovering helpful reviews to save reader time. Unlike existing works, our paper targets the company domain, where understanding a company heavily relies on knowledge of uncommon attributes and quantitative information, as indicated by expert-written reviews. Therefore, our definition of salient facts serves as another dimension to analyze massive reviews, and our work complements existing efforts towards mining the most useful information from reviews.

Supervised learning: existing works mostly adopt supervised learning when developing automatic extractors because supervised models can automatically learn to differentiate positive and negative instances from human labels. There are three popular categories of supervised models, depending on input sequence representation: word occurrence models [12,13,15], such as logistic regression [25] and support vector machine [26], representing a text as a bag of words and thus suffering from limited vocabulary when the number of training data is small. Word vector models [8,13,14,17,27,28],such as convolutional neural networks [29] and long short-term memory [30], represent a text as a matrix of word embeddings and can thereby process unseen words through their embeddings. Recently, pretrained models [8,9], such as BERT [5], ALBERT [6], and RoBERTa [7], have emerged representing a text as a high-dimensional vector by aggregating word embeddings. Due to the high dimension (e.g., 768 in BERT) and large-scale parameters (e.g., 110M in BERT) for aggregation, pretrained models appear to be the most promising solutions for extractions. In fact, among all different models, pretrained models achieved the best F1 scores and are thus the base models for our work.

Label scarcity: the problem of salient fact extraction falls into the big category of text classification. However, the unique challenge here is label sparsity. The ratios of salient facts in raw reviews are extremely low (<10%) due to the nature of uncommon attributes and quantitative descriptions that require solid domain-specific knowledge from crowd reviewers. As a result, collecting a large number of salient facts for model training is very difficult. We thus propose a label propagation method to expand existing salient facts with two benefits. First, the method expands the input tokens of a input sentence towards instructing pretrained models about whether the input carries uncommon attributes or quantitative descriptions. Second, the method fetches more salient fact instances from the ample unlabeled corpus to enable pretrained models seeing more salient facts. The label propagation method was specifically designed to suit the nature of uncommon attributes and quantitative information, and is thus complementary to existing techniques such as data augmentation [31–33] and active learning [34–36]. A combination of existing techniques can further improve extraction quality. However, it is nontrivial to adapt existing techniques here due to increased algorithmic complexity; therefore, incorporating existing techniques is a fascinating future direction for this work.

7. Extraction

In this section, we present extracted salient facts for qualitative analysis. We used BERT as the representative pretrained models. We also present extractions using existing solutions.

7.1. Extraction Comparison

We present salient facts extracted from reviews about Google on Table 9. We also present salient facts extracted by baseline algorithms TextRank, K-means, Longest, and Random. TextRank [37] formulates sentences and their similarity relation into a graph, and extracts texts with the highest PageRank weights. K-means clusters sentences into a number of centroids and extracts the centroid sentences. Longest chooses the longest sentence from the corpus. Random randomly selects sentences from the corpus. These algorithms form a complete set of existing solutions for mining informative texts from a large corpus.

Table 9. Extractions of various methods on Google dataset with attributes and descriptions marked in red and blue, respectively. Our extractions revealed finer-grained attributes (see red) and distilled numeric knowledge (see blue).

Method	Extractions
Ours	on campus laundry rooms, lots of gyms, cars on demand in case you have to drive during the day. Flexible working hours, 90% of health insurance paid for, 12 weeks paid parental leave as a secondary care giver, free breakfast/lunch/dinner.
TextRank	lots of happy hours and the free food is as great as everyone says it is. Solving challenging and interesting problems that matter to people.
Kmeans	Interesting work. Google.
Longest	Chapter 4 of "English to Go" deals with Aeon, one of the other mega English teaching companies, and is entitled "Aeon's Cult of Impersonality." An earlier chapter, chapter 2, that deals specifically with NOVA doesn't delve into the cultlike training...
Random	free food. awesome place to work, great salary, smart people.

Finer-grained attribute discovery: extraction examples show that our method extracted salient facts that contained finer-grained attributes than those extracted by the baseline methods. Representative attributes include laundry room, gyms, cars, and museum tickets. These attributes describe concrete properties about the company and are less common in the company domain. In contrast, extractions by the existing solutions tend to contain common attributes such as food, problem, work, salary, or people, which are popular and general topics about companies. The extractions by Longest did not reveal company attributes since the method retrieves long yet fake reviews that are copies of external literature. Results suggest that salient facts are informative when presenting specific or unique attributes of a company to readers.

Numeric knowledge distillation: our extractions distill numeric knowledge compared with extractions from existing solutions. Representative knowledge includes 90% paid health insurance, 12 weeks paid parental leave, and free meals provided by the company. Knowledge is objective since it quantitatively describes attributes. In contrast, extractions from existing solutions mostly use subjective descriptions such as "lots of", "great", and "awesome". These subjective descriptions are biased towards reviewers. Results suggest that salient facts can provide unbiased and reliable descriptions to readers.

7.2. Expert Comment Recognition

Online comments are written by different people. Some writers with better knowledge about entities tend to give comments that are more informative. We refer to such writers as experts, and their comments as expert comments. In order to show the most informative comments to readers, a salient fact extractor should rank expert comments higher than other comments.

To understand whether our trained model could rank expert comments higher, we curated a collection of comments from online company reviews and FutureFuel. Online comments are those that we labeled as nonsalient (some representatives are in Table 3) and were thereby treated as nonexpert reviews. FutureFuel comments are those that came from invited writers and were thereby treated as expert reviews. We then sorted the collection of nonexpert and expert comments by prediction scores in descending order. A higher prediction score indicated a higher probability to be an expert comment.

Ranking results of Google and Amazon datasets are shown in Table 10. In the optimal case, all comments in the top-k list were expert comments. We show the number of expert comments of our model and a baseline that randomly shuffles all comments. Our model consistently achieved better results than the baseline in both the Google and the Amazon dataset, as shown in Table 10. In top 4 lists, all comments returned by our models were expert comments. In top 10 lists, 9 comments were expert comments in both Google and Amazon. Results indicate that our model could identify expert comments with nearly 100% accuracy. In the collection or comments that came from different people, our models could effectively recognize comments that had been written by experts, could and this ensure that readers are shown the most informative contents.

Table 10. Number of expert comments in top list after sorting all comments by prediction scores. Baseline randomly shuffles all comments.

Google (14 Expert Comments + 14 Online Comments)		
Top List	**Ours**	**Baseline**
Top 4	4	2
Top 10	9	5
Top 14	13	7
Amazon (16 Expert Comments + 16 Online Comments)		
Top List	**Ours**	**Baseline**
Top 4	4	2
Top 10	9	5
Top 16	15	8

8. Conclusions

In this paper, we proposed a task of extracting salient facts from online company reviews. In contrast to reviews written by experts, only a few online reviews contain useful and salient information about a particular company, which creates a situation where the solution can only rely on highly skewed and scarce training data. To address the data scarcity issue, we developed two data enrichment methods, (1) representation enrichment and (2) label propagation, to boost the performance of supervised learning models. Experimental results showed that our data enrichment methods could successfully help in training a high-quality salient fact extraction model with fewer human annotations.

Author Contributions: Conceptualization, J.L., N.B., A.W., C.-Y.H., E.H. and Y.S.; methodology, J.L., N.B., A.W., E.H. and Y.S.; software, J.L., C.-Y.H.; validation, J.L., N.B., E.H. and Y.S.; formal analysis, J.L., N.B., E.H. and Y.S.; investigation, J.L., N.B., E.H. and Y.S.; resources, Y.S.; data curation, J.L., N.B. and Y.S.; writing—original draft preparation, J.L., N.B. and Y.S.; writing—review and editing, J.L., N.B., C.-Y.H., E.H. and Y.S.; visualization, J.L., N.B.; supervision, E.H. and Y.S.; project administration, J.L. and E.H.; All authors have read and agreed to the published version of the manuscript.

Funding: This research received no external funding.

Institutional Review Board Statement: Not applicable.

Informed Consent Statement: Not applicable.

Data Availability Statement: https://github.com/rit-git/tagging/tree/master/data.

Conflicts of Interest: The authors declare no conflict of interest.

References

1. Local Consumer Review Survey. 2019. Available online: https://www.brightlocal.com/research/local-consumer-review-survey/ (accessed on 22 August 2020)
2. Business Insider Google Perks. 2017. Available online: https://www.businessinsider.com/google-employee-best-perks-benefits-2017-11 (accessed on 22 August 2020)
3. Business Insider Amazon Perks. 2018. Available online: https://www.businessinsider.com.au/amazon-hq2-employee-perks-2018-1 (accessed on 22 August 2020)
4. FutureFuel Employee Benefits Summary. 2020. Available online: https://futurefuel.io/employee-benefits/ (accessed on 22 August 2020)
5. Devlin, J.; Chang, M.; Lee, K.; Toutanova, K. BERT: Pre-training of Deep Bidirectional Transformers for Language Understanding. In Proceedings of the 2019 Conference of the North American Chapter of the Association for Computational Linguistics: Human Language Technologies, NAACL-HLT 2019, Minneapolis, MN, USA, 2–7 June 2019; Volume 1, pp. 4171–4186.
6. Lan, Z.; Chen, M.; Goodman, S.; Gimpel, K.; Sharma, P.; Soricut, R. ALBERT: A Lite BERT for Self-supervised Learning of Language. *arXiv* **2020**, arXiv:1909.11942.
7. Liu, Y.; Ott, M.; Goyal, N.; Du, J.; Joshi, M.; Chen, D.; Levy, O.; Lewis, M.; Zettlemoyer, L.; Stoyanov, V. RoBERTa: A Robustly Optimized BERT Pretraining Approach. *arXiv* **2019**, arXiv:1907.11692.
8. Negi, S.; Daudert, T.; Buitelaar, P. SemEval-2019 Task 9: Suggestion Mining from Online Reviews and Forums. In Proceedings of the 13th International Workshop on Semantic Evaluation, SemEval@NAACL-HLT 2019, Minneapolis, MN, USA, 6–7 June 2019; pp. 877–887.
9. Liu, J.; Wang, S.; Sun, Y. OleNet at SemEval-2019 Task 9: BERT based Multi-Perspective Models for Suggestion Mining. In Proceedings of the 13th International Workshop on Semantic Evaluation, SemEval@NAACL-HLT 2019, Minneapolis, MN, USA, 6–7 June 2019; pp. 1231–1236.
10. Xie, Q.; Dai, Z.; Hovy, E.H.; Luong, T.; Le, Q. Unsupervised Data Augmentation for Consistency Training. *Adv. Neural Inf. Process. Syst.* **2020**, *33*, 6256–6268.
11. Natural Language Tookit. 2020. Available online: https://www.nltk.org/ (accessed on 22 August 2020).
12. Guy, I.; Mejer, A.; Nus, A.; Raiber, F. Extracting and Ranking Travel Tips from User-Generated Reviews. In Proceedings of the 26th International Conference on World Wide Web, WWW 2017, Perth, Australia, 3–7 April 2017; pp. 987–996.
13. Negi, S.; Buitelaar, P. Towards the Extraction of Customer-to-Customer Suggestions from Reviews. In Proceedings of the 2015 Conference on Empirical Methods in Natural Language Processing, EMNLP 2015, Lisbon, Portugal, 17–21 September 2015; pp. 2159–2167.
14. Wang, S.; Phan, N.; Wang, Y.; Zhao, Y. Extracting API Tips from Developer Question and Answer Websites. In Proceedings of the 16th International Conference on Mining Software Repositories, MSR 2019, Montreal, QC, Canada, 26–27 May 2019; pp. 321–332
15. Chen, N.; Lin, J.; Hoi, S.C.H.; Xiao, X.; Zhang, B. AR-Miner: Mining Informative Reviews for Developers from Mobile App Marketplace. In Proceedings of the 36th International Conference on Software Engineering, ICSE '14, Hyderabad, India, 31 May–7 June 2014; pp. 767–778.
16. Hua, X.; Nikolov, M.; Badugu, N.; Wang, L. Argument Mining for Understanding Peer Reviews. In Proceedings of the 2019 Conference of the North American Chapter of the Association for Computational Linguistics: Human Language Technologies, NAACL-HLT 2019, Minneapolis, MN, USA, 2–7 June 2019; Volume 1, pp. 2131–2137.

17. Novgorodov, S.; Elad, G.; Guy, I.; Radinsky, K. Generating Product Descriptions from User Reviews. In Proceedings of the World Wide Web Conference, WWW 2019, San Francisco, CA, USA, 13–17 May 2019; pp. 1354–1364.
18. Elad, G.; Guy, I.; Novgorodov, S.; Kimelfeld, B.; Radinsky, K. Learning to Generate Personalized Product Descriptions. In Proceedings of the 28th ACM International Conference on Information and Knowledge Management, CIKM 2019, Beijing, China, 3–7 November 2019; pp. 389–398.
19. Zhang, X.; Qiao, Z.; Ahuja, A.; Fan, W.; Fox, E.A.; Reddy, C.K. Discovering Product Defects and Solutions from Online User Generated Contents. In Proceedings of the World Wide Web Conference, WWW 2019, San Francisco, CA, USA, 13–17 May 2019; pp. 3441–3447.
20. Morales, A.; Zhai, C. Identifying Humor in Reviews using Background Text Sources. In Proceedings of the 2017 Conference on Empirical Methods in Natural Language Processing, EMNLP 2017, Copenhagen, Denmark, 9–11 September 2017; pp. 492–501.
21. Zhang, X.; Zhao, J.J.; LeCun, Y. Character-level Convolutional Networks for Text Classification. In Proceedings of the Advances in Neural Information Processing Systems 28: Annual Conference on Neural Information Processing Systems 2015, Montreal, QC, Canada, 7–12 December 2015; pp. 649–657.
22. Yelp Dataset Challenge. 2020. Available online: https://www.yelp.com/dataset/documentation/main (accessed on 22 August 2020)
23. O'Mahony, M.P.; Smyth, B. Learning to Recommend Helpful Hotel Reviews. In Proceedings of the 2009 ACM Conference on Recommender Systems, RecSys 2009, New York, NY, USA, 23–25 October 2009; pp. 305–308.
24. Lee, P.; Hu, Y.; Lu, K. Assessing the helpfulness of online hotel reviews: A classification-based approach. *Telemat. Informat.* **2018**, *35*, 436–445. [CrossRef]
25. Friedman, J.; Hastie, T.; Tibshirani, R. *The Elements of Statistical Learning*; Springer: Cham, Switzerland, 2001.
26. Suykens, J.A.K.; Vandewalle, J. Least Squares Support Vector Machine Classifiers. *Neural Process. Lett.* **1999**, *9*, 293–300. [CrossRef]
27. Gao, C.; Zeng, J.; Lyu, M.R.; King, I. Online App Review Analysis for Identifying Emerging Issues. In Proceedings of the 40th International Conference on Software Engineering, ICSE 2018, Gothenburg, Sweden, 27 May–3 June 2018; pp. 48–58.
28. Gao, C.; Zheng, W.; Deng, Y.; Lo, D.; Zeng, J.; Lyu, M.R.; King, I. Emerging app issue identification from user feedback: Experience on WeChat. In Proceedings of the 41st International Conference on Software Engineering: Software Engineering in Practice, ICSE (SEIP) 2019, Montreal, QC, Canada, 25–31 May 2019; pp. 279–288.
29. Kim, Y. Convolutional Neural Networks for Sentence Classification. In Proceedings of the 2014 Conference on Empirical Methods in Natural Language Processing, EMNLP 2014, Doha, Qatar, 25–29 October 2014; pp. 1746–1751.
30. Hochreiter, S.; Schmidhuber, J. Long Short-Term Memory. *Neural Comput.* **1997**, *9*, 1735–1780. [CrossRef] [PubMed]
31. Gao, F.; Zhu, J.; Wu, L.; Xia, Y.; Qin, T.; Cheng, X.; Zhou, W.; Liu, T. Soft Contextual Data Augmentation for Neural Machine Translation. In Proceedings of the 57th Conference of the Association for Computational Linguistics, ACL 2019, Florence, Italy, 28 July–2 August 2019; Volume 1, pp. 5539–5544.
32. Wei, J.W.; Zou, K. EDA: Easy Data Augmentation Techniques for Boosting Performance on Text Classification Tasks. In Proceedings of the 2019 Conference on Empirical Methods in Natural Language Processing and the 9th International Joint Conference on Natural Language Processing, EMNLP-IJCNLP 2019, Hong Kong, China, 3–7 November 2019; pp. 6381–6387.
33. Rizos, G.; Hemker, K.; Schuller, B.W. Augment to Prevent: Short-Text Data Augmentation in Deep Learning for Hate-Speech Classification. In Proceedings of the 28th ACM International Conference on Information and Knowledge Management, CIKM 2019, Beijing, China, 3–7 November 2019; pp. 991–1000.
34. McCallum, A.; Nigam, K. Employing EM and Pool-Based Active Learning for Text Classification. In Proceedings of the Fifteenth International Conference on Machine Learning (ICML 1998), Madison, WI, USA, 24–27 July 1998; pp. 350–358.
35. Yan, Y.; Huang, S.; Chen, S.; Liao, M.; Xu, J. Active Learning with Query Generation for Cost-Effective Text Classification. In Proceedings of the Thirty-Fourth AAAI Conference on Artificial Intelligence, AAAI, New York, NY, USA, 7–12 February 2020; pp. 6583–6590.
36. Cormack, G.V.; Grossman, M.R. Scalability of Continuous Active Learning for Reliable High-Recall Text Classification. In Proceedings of the 25th ACM International Conference on Information and Knowledge Management, CIKM 2016, Indianapolis, IN, USA, 24–28 October 2016; pp. 1039–1048.
37. Mihalcea, R.; Tarau, P. TextRank: Bringing Order into Text. In Proceedings of the 2004 Conference on Empirical Methods in Natural Language Processing , EMNLP 2004, A meeting of SIGDAT, a Special Interest Group of the ACL, held in conjunction with ACL 2004, Barcelona, Spain, 25–26 July 2004; pp. 404–411.

Proceeding Paper

Long-Tail Zero and Few-Shot Learning via Contrastive Pretraining on and for Small Data †

Nils Rethmeier [1,2,*], Isabelle Augenstein [2]

1 German Research Center for AI, SLT-Lab, Alt-Moabit 91c, 10559 Berlin, Germany
2 Department of Computer Science, Copenhagen University, Universitetsparken 1, 2100 Copenhagen, Denmark, augenstein@di.ku.dk
* Correspondence: nils.rethmeier@dfki.de

† Presented at the AAAI Workshop on Artificial Intelligence with Biased or Scarce Data (AIBSD), Online, 28 February 2022.

Abstract: Preserving long-tail, minority information during model compression has been linked to algorithmic fairness considerations. However, this assumes that large models capture long-tail information and smaller ones do not, which raises two questions. One, how well do large pretrained language models encode long-tail information? Two, how can small language models be made to better capture long-tail information, without requiring a compression step? First, we study the performance of pretrained Transformers on a challenging new long-tail, web text classification task. Second, to train small long-tail capture models we propose a contrastive training objective that unifies self-supervised pretraining, and supervised long-tail fine-tuning, which markedly increases tail data-efficiency and tail prediction performance. Third, we analyze the resulting long-tail learning capabilities under zero-shot, few-shot and full supervision conditions, and study the performance impact of model size and self-supervision signal amount. We find that large pretrained language models do not guarantee long-tail retention and that much smaller, contrastively pretrained models better retain long-tail information while gaining data and compute efficiency. This demonstrates that model compression may not be the go-to method for obtaining good long-tail performance from compact models.

Keywords: contrastive language models; long-tail compression, text-to-text; self-supervised contrastive pretraining, contrastive autoencoder.

Citation: Rethmeier, N.; Augenstein, I Long-Tail Zero and Few-Shot Learning via Contrastive Pretraining on and for Small Data. *CSFM* **2022**, 3, 10. https://doi.org/10.3390/cmsf2022003010

Academic Editors: Kuan-Chuan Peng and Ziyan Wu

Published: 20 May 2022

1. Introduction

Long-tail information has been found to be disproportionately affected during model compression, which has in turn been linked to reducing aspects of algorithmic fairness for minority information [1,2]. Additionally, real-world data is subject to long-tail learning challenges such as imbalances, few-shot learning, open-set recognition [3], or feature and label noise [4,5]. Crucially, works by Hooker et al. [6], Zhuang et al. [7] find that common long-tail evaluation measures like top-k metrics mask tail prediction performance losses. Current works on long-tail preservation in smaller models are focused on compressing large, supervised computer vision models [3,8–11], while general long-tail learning methods only study supervised contrastive learning.

In this work, we extend the field of 'long-tail preservation in compact models' to (self-supervised) *pretrained language models* (PLMs), and investigate whether contrastive language modeling (CLM) can be used to train a small, long-tail preserving model which does not require compression or large pretrained models. In this context, large PLMs are an important point of reference since they are often assumed to be base models for use in arbitrary NLP downstream tasks, as a trade-off for their large pretraining costs. These models are pretrained over many text domains in the hopes of achieving partial in-domain pretraining that later overlaps with arbitrary downstream applications. This works well except in cases where fine-tuning data is limited [12]. Unfortunately, training data and sub-domains in the tail of a distribution are always limited and diverse by definition, which foreseeably increases the domain distribution mismatch between large PLMs and long-tail distributed end-task data. Hence, in order to train long-tail preserving models, it is useful to study small-scale, but in-domain pretraining, which ideally, is similarly or more compute efficient than fine-tuning a large PLM, while still achieving superior long-tail prediction performance. Thus, we first evaluate a large PLM in a challenging long-tail tag prediction setup (see Section 4) and then move on to propose a small contrastive language model (CLM) to answer the following three research questions.

- RQ-1: Does a large pretrained language model, in this case, RoBERTa [13], achieve good long-tail class prediction performance (Section 5.1)?
- RQ-2: Can we extend language models such that a small language model can retain accurate long-tail information, with overall training that is computationally cheaper than fine-tuning RoBERTa?
- RQ-3: What are the long-tail prediction performance benefits of small CLMs that unify self-supervised and supervised contrastive learning?

Contributions

We address RQ-2 by proposing a contrastive language model objective that *unifies supervised learning with self-supervised pretraining* to produce a *small model, with strong long-tail retention* that is cheap to compute, thereby avoiding the need for compressing a large model. This takes inspiration from supervised contrastive learning, which is known to improve long-tail learning in NLP [8,14,15]. However, we add *self-supervised contrastive learning* since its effect has not been studied in the context of language models for long-tail learning, especially not with the requirement of producing small models. We call this unified learning objective: Contrastive Long-tail Efficient Self-Supervision or CLESS. The method constructs pseudo-labels from input text tokens to use them for contrastive self-supervised pretraining. During supervised fine-tuning on real (long-tail) labels, the model directly reuses the self-supervision task head to predict real, human-annotated, text labels. Thus, we unify self-supervised and supervised learning regimes into a 'text-to-text' approach. This builds on ideas for large PLMs that use 'text-to-text' prediction like T5 [16] and extends them to contrastive self-supervision to ensure long-tail retention in small language models that pretrain efficiently, even under strong data limitations. Using a 'text-to-text' prediction objective allows for modeling arbitrary NLP tasks by design, though in this work we focus exclusively on improving the under-studied field of long-tail language modeling. We evaluate RQ-1 and RQ-2 by comparing RoBERTa against CLESS regarding long-tail prediction in Section 5.1. To address RQ-3, we study three long-tail learning performance aspects. (RQ-3.1) We study how well our contrastive self-supervised pretraining generalizes to long-tail label prediction without using labeled examples, i.e. zero-shot, long-tail prediction in Section 5.2. (RQ-3.2) We evaluate how zero-shot performance is impacted by increased model size and pseudo-label amount during self-supervised pretraining (Section 5.2). (RQ-3.3) Finally, we investigate our models' few-shot learning capabilities during supervised long-tail fine-tuning and compare the results to the RoBERTa model in Section 5.3.

2. Related Work

In this section, we summarize related work and how it influenced our method design and evaluation strategy decisions.

2.1. Long-Tail Compression

Works by Hooker et al. [1,6] raised awareness of the disproportionate loss of long-tail information during model compression and the undesirable rise in algorithmic bias and fairness issues this may cause. Other works such as Liu et al. [3] pointed out that real-world learning is always long-tailed and that few-shot and zero-shot learning settings naturally arise in tailed, real-world distributions. To make matters worse, real-world long-tail data is highly vulnerable to noise, which creates drastic learning and evaluation challenges, especially for self-supervised learning methods. For example, D'souza et al. [4] identify types of noise that especially impact long-tail data prediction and Zhuang et al. [7] find that noise disproportionately affects long-tail metrics. In fact, all the aforementioned show that top-k metrics hide long-tail performances losses. This means that we need long-tail sensitive evaluation, which inspired us to use Average Precision as a measure. In addition, we split tail analysis into 5 buckets that all contain an equal amount of positive labels, where each bucket contains increasingly more and rarer classes—see Section 4. These label imbalances in long-tail tasks make manual noise treatment very cumbersome, but fortunately, contrastive objectives are naturally robust to label noise as we will detail in the paragraph below.

2.2. Contrastive Learning Benefits

Contrastive objectives like Noise Contrastive Estimation (NCE), have been shown to be much more robust against label noise overfitting than the standard cross-entropy loss [17]. Additionally, Zimmermann et al. [18] found that contrastive losses can "recover the true data distribution even from very limited learning samples". Supervised contrastive learning methods like Chang et al. [8], Liu et al. [14], Pappas and Henderson [15], Zhang et al. [19] have repeatedly demonstrated improved long-tail learning. Finally, Jiang et al. [11] recently proposed contrastive long-tail compression into smaller models. However, this still leaves the research question (RQ-1), whether large models learn long-tail well enough in the first place, unanswered. These observations, learning properties and open research questions inspired us to forgo large model training and the subsequent compression by instead training small contrastive models and extending them with contrastive self-supervision to combine the benefits of language model pretraining and contrastive learning. This imbues a small (contrastive language) model with strong long-tail retention capabilities, as well as with data-efficient learning for better zero to few-shot learning—as is detailed in the results Section 5.

2.3. Long-Tail Learning

Long-tail learning has prolific subfields like extreme classification, which is concerned with supervised long-tail learning and top-line metric evaluation. The field provides varied approaches for different data input types like images [3], categorical data, or text classification using small supervised [14] or large supervision fine-tuned PLMs like Chang et al. [8] for supervised tail learning. However, these methods only explore *supervised* contrastive learning and limit their evaluation to *top-line metrics*, which, as mentioned above, mask long-tail performance losses. This naturally leads us to explore the effects of *self-supervised contrastive* learning (or pretraining) as one might expect such pretraining to enrich long-tail information before tail learning supervision. Additionally, as mentioned above, we use Average Precision over all classes, rather than top-k class, to *unmask long-tail performance losses*.

2.4. Negative and Positive Generation

As surveys like Musgrave et al. [20], Rethmeier and Augenstein [21] point out, traditional contrastive learning research focuses on generating highly informative (hard) **negative samples**, since most contrastive learning objectives only use *a single positive learning sample* and *b* (bad) negative samples—Musgrave et al. [20] give an excellent overview. However, if too many negative samples are generated they can collide with positive samples, which degrades learning performance [22]. More recent computer vision works like Khosla et al. [23], Ostendorff et al. [24] propose generating multiple **positive** samples to boost *supervised contrastive learning* performance, while Wang and Isola [25] show that, when generating positive samples, the representations of positives should be close (related) to each other. Our method builds on these insights and extends them to *self-supervised* *contrastive learning* and to the language model domain using a straightforward extension to NCE. Instead of using only one positive example the like standard NCE by Mnih and Teh [26], our method uses g good (positive) samples (see Section 3). To ensure that positive samples are representationally close (related) during self-supervised contrastive pretraining, we use words from a current input text as positive 'pseudo-labels'—i.e., we draw self-supervision pseudo-labels from a related context. Negative pseudo-labels (words) are drawn as words from other in-batch text inputs, where negative sample words are not allowed not appear in the current text to avoid the above-mentioned collision of positive and negative samples.

2.5. Data and Parameter Efficiency

Using CNN layers can improve data and compute efficiency over self-attention layers as found by various works [27–29]. data-efficiency is paramount when pretraining while data is limited, which, for (rare) long-tail information, is by definition, always the case. Radford et al. [30] find that replacing a Transformer language encoder with a CNN backbone increases zero-shot data-efficiency 3 fold. We thus use a small CNN text encoder, while for more data abundant or short-tail pretraining scenarios a self-attention encoder may be used instead. **Our method is designed to increase self-supervision signal, i.e., by sampling more positive and negatives, to compensate for a lack of large pretraining data (signal)—since rare and long-tailed data is always limited**. It is our goal to skip compression and still train small, long-tail prediction capable models. Notably, CLESS pretraining does not require special learning rate schedules, residuals, normalization, warm-ups, or a modified optimizer as do many BERT variations [13,31,32].

2.6. Label Denoising

Label dropout of discrete $\{0,1\}$ labels has been shown to increase label noise robustness by [33]. We use dropout on both the dense text and label embeddings. This creates a 'soft', but dense label noise during both self-supervised and supervised training, which is also similar to sentence similarity pretraining by Gao et al. [34], who used text embedding dropout rather than label embedding dropout to generate augmentations for contrastive learning.

3. CLESS: Unified Contrastive Self-supervised to Supervised Training and Inference

As done in natural language usage, we express labels as words, or more specifically as word embeddings, rather than as $\{0,1\}$ label vectors. CLESS then learns to contrastively (mis-)match <text embedding, (pseudo/real) label embedding> pairs as overviewed in Figure 1. For self-supervised pretraining, we in-batch sample g (good) positive and b (bad) negative <text, pseudo label> embedding pairs per text instance to then learn good and bad matches from them. Positive pseudo labels are a sampled subset of words that appear in the current text instance. Negative pseudo labels are words sampled from the other texts within a batch. Crucially, negative words (pseudo labels) can not be the same words as positive words (pseudo labels)—i.e. $\mathbf{w}_i^+ \cap \mathbf{w}_j^- = \emptyset$.

This deceptively simple sampling strategy ensures that we fulfill two important criteria for successful *self-supervised contrastive learning*. One, using multiple positive labels improves learning if we draw them from a similar (related) context, as Wang and Isola [25] proved. Two, we avoid collisions between positive and negative samples, which otherwise degrades learning when using more negatives as Saunshi et al. [22] find. Similarly, for supervised learning, we use g positive, real labels and undersample b negative labels to construct <text, positive/negative real label> pairs. A text-2-label classifier ⑤ learns to match <text, label> embedding pairs using a noise contrastive loss [35], which we extend to use g positives rather than just one. This unifies self-supervised and supervised learning as contrastive 'text embedding to (label) text embedding matching' and allows direct transfer like zero-shot predictions of real labels after pseudo label pretraining—i.e. without prior training on other real labels as required by methods like [15,19,36]. Below, we describe our approach and link specific design choices to insights from existing research in steps ①-⑥.

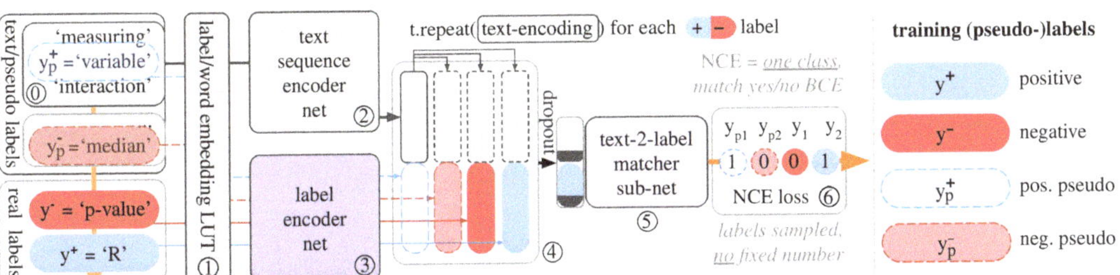

Figure 1. Contrastive <text, pseudo/real label> embedding pair matcher model: A word embedding layer E ① embeds text and real/pseudo labels, where labels are word IDs. CLESS embeds a text ('measuring variable interaction'), real positive (R) or negative (p-value) labels, and positive (variable) or negative (median) pseudo labels. A sequence encoder T ② embeds a single text, while a label encoder L ③ embeds c labels. Each text has multiple (pseudo) labels, so the text encoding $\mathbf{t}_i$ is repeated for, and concatenated with, each label encoding $\mathbf{l}^{\circ}_{i,l}$. The resulting batch of <text embedding, label embedding> pairs $[[\mathbf{t}_i, \mathbf{l}^{\circ}_{i,1}], \ldots, [\mathbf{t}_i, \mathbf{l}^{\circ}_{i,c}]]$ ④ are fed into a 'matcher' classifier ⑤ that is trained in ⑥ as a binary noise contrastive estimation loss L_B [35] over multiple label (mis-)matches $\{0,1\}$ per text instance $\mathbf{t}_i$. Unlike older works, we add contrastive self-supervision over pseudo labels as a pretraining mechanism. Here, the word 'variable' is a positive self-supervision (pseudo) label for a text instance $\mathbf{t}_i$, while words from other in-batch texts, e.g. 'median', provide negative pseudo labels.

We give the model a text instance i of words $\mathbf{w}_i$ and a set of positive and negative label words $\mathbf{w}_i^{\circ} = \mathbf{w}_i^{+} \oplus \mathbf{w}_j^{-} \in \mathbb{R}^{c=g+b}$. We also construct a label indicator $\mathbb{I}_i$ as ground truth labels for the binary NCE loss in ⑥. This label indicator contains a g-sized vector of ones $\mathbf{1} \in \mathbb{N}_0^g$ to indicate positive (matching) <text, label> embedding pairs and a b-sized zero vector $\mathbf{0} \in \mathbb{N}_0^b$ to indicated mismatching pairs, resulting in the indicator

$$\mathbb{I}_i = \{\mathbf{1} \oplus \mathbf{0}\} \in \mathbb{N}_0^{c=g+b} \qquad ⓪$$

CLESS then encodes input text and labels in three steps ①-③. First, both the input text (words) $\mathbf{w}_i$ and the labels $\mathbf{w}_i^{\circ}$ are passed through a shared embedding layer ① to produce $E(\mathbf{w}_i)$ as text embeddings and $E(\mathbf{w}_i^{\circ})$ as label embeddings. Then, the text embeddings are encoded via a text encoder T ②, while labels are encoded by a label encoder L as follows:

$$E(\mathbf{w}_i), E(\mathbf{w}_i^{\circ}) \qquad ①$$

$$\mathbf{t}_i = T(E(\mathbf{w}_i)) \qquad ②$$

$$\mathbf{L}_i^{\circ} = L(E(\mathbf{w}_i^{\circ})) = [\mathbf{l}_{i,1}^{+}, \ldots, \mathbf{l}_{i,g}^{+}, \mathbf{l}_{i,1}^{-}, \ldots, \mathbf{l}_{i,b}^{-}] \qquad ③$$

To make model learning more data-efficient we initialize the embedding layer E with fastText word embeddings that we train on the 60MB of *in-domain text data*. Such word embedding training only computes a few seconds, while enabling one to make the text encoder architecture small, but well initialized. The text encoder T consists of a single, k-max-pooled CNN layer followed by a fully connected layer for computation speed and data-efficiency [30,37,38]. As a label encoder L, we average the embeddings of words in a label and feed them through a fully connected layer—e.g. to encode a label 'p-value' we simply calculate the mean word embedding for the words 'p' and 'value'.

To learn whether a text instance embedding $\mathbf{t}_i$ matches any of the c label embeddings $\mathbf{l}_{i,\cdot}^{\circ} \in \mathbf{L}_i^{\circ}$, we repeat the text embedding $\mathbf{t}_i$, c times, and concatenate text and label embeddings to get a matrix $\mathbf{M}_i$ of <text, label> embedding pairs:

$$\mathbf{M}_i = [[\mathbf{t}_i, \mathbf{l}_{i,1}^{+}], \ldots, [\mathbf{t}_i, \mathbf{l}_{i,c}^{-}]] \quad ④$$

This text-label paring matrix $\mathbf{M}_i$ is then passed to the matcher network M ⑤, which first applies dropout to each text-label embedding pair and then uses a three layer MLP to produce a batch of c label match probabilities:

$$\mathbf{p}_i = \{\sigma(M(\mathbf{M}_{i,1})), \ldots, \sigma(M(\mathbf{M}_{i,c}))\} \quad ⑤$$

Here, applying dropout to label and text embeddings induces a *dense version of label noise*. Discrete {0,1} label dropout has been shown to improve robustness to label noise in Szegedy et al. [33], Lukasik et al. [39]. Because we always predict correct pseudo labels in pretraining, this forces the classifier to learn to correct dropout induced label noise.

Finally, we use a binary noise contrastive estimation loss as in [35], but extend it to use g positives, not one.

$$L_B = -\frac{1}{c}\sum_{l=1}^{g+b=c} \mathbb{I}_{i,l} \cdot log(\mathbf{p}_{i,l}) + (1 - \mathbb{I}_{i,l}) \cdot log(1 - \mathbf{p}_{i,l}) \quad ⑥$$

Here, L_B is the mean binary cross-entropy loss of g positive and b negative labels—i.e. it predicts $c = b+g$ label probabilities $\mathbf{p}_i$, where the label indicators $\mathbb{I}_i$ from ① are used as ground truth labels.

Though we focus on evaluating CLESS for long-tail prediction in this work, other NLP tasks such as question answering or recognizing textual entailment can similarly be modeled as contrast pairs $<X = \text{'}text\ 1\ [sep]\ text\ 2\text{'}, Y = \text{'}is\ answer\text{'}>$. Unlike T5 language models [16], this avoids translating back and forth between discrete words and dense token embeddings. Not using T5s' softmax objective, also allows for predicting unforeseen (unlimited) test classes (label). We provide details on hyperparameter tuning of CLESS for self-supervised and supervised learning in Appendix C.

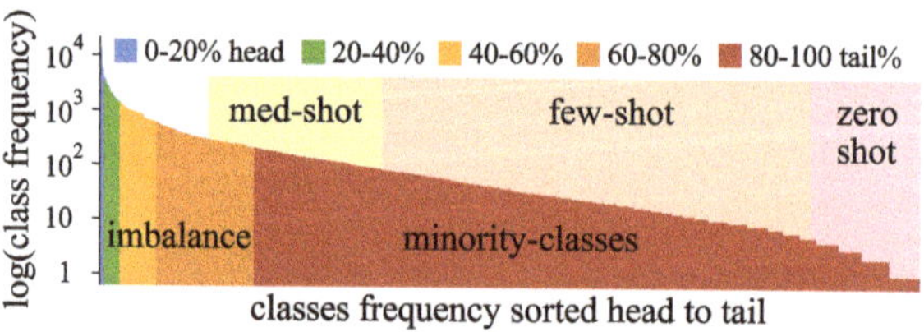

Figure 2. Head to long-tail as 5 balanced class bins: We bin classes by label frequency. Each bin contains equally many active label occurrences. Classes within a bin are imbalanced and become few-shot or zero-shot towards the tail, especially after train/dev/test splitting. Class frequencies are given in log scale—task data details in Section 4.

4. Data: Resource Constrained, Long-Tail, Multi-Label, Tag Prediction

To study efficient, small model, long-tail learning for 'text-to-text' pretraining models, we choose a multi-label question tag prediction dataset as a testbed. We use the "Questions from Cross Validated" dataset, where machine learning concepts are tagged per question—https://www.kaggle.com/stackoverflow/statsquestions, accessed on 30 August 2021. This dataset is small (80MB of text), and entails solving a challenging 'text-to-text' long-tailed prediction task. The dataset has 85k questions with 244k positive labels, while we do not use answer texts. As with many real-world problems, labels are vague, since tagging was crowd-sourced. This means that determining the correct amount of tags per question (label density) is hard, even for humans. The task currently has no prior state-of-the-art. As seen in Figure 2, the datasets' class occurrence frequencies are highly long-tailed, i.e. the 20% most frequently occurring classes result in 7 'head' classes, while the 20% least frequent (rightmost) label occurrences cover 80% or 1061/1315 of classes. Tags are highly sparse—at most 4 out of 1315 tags are labeled per question. We pretrain fastText word embeddings on the unlabeled text data to increase learning efficiency, and because fastText embeddings only take a few seconds to pretrain. The full details regarding preprocessing can be found in Appendix A.

Long-tail evaluation metrics and challenges:

Long-tail, multi-label classification is challenging to evaluate because (i) top-k quality measures mask performance losses on long-tailed minority classes as Hooker et al. [6] point out. Furthermore, (ii) measures like ROC_{AUC} overestimate performance under class imbalance [40,41], and (iii) discrete measures like F-score are not scalable, as they require discretization threshold search under class imbalance. Fortunately, the Average Precision score $AP = \sum_n (R_n - R_{n-1})P_n$ addresses issues (i-iii), where P_n and R_n are precision and recall at the nth threshold. We choose AP_{micro} weighting as this score variant is the hardest to improve.

5. Results

In this section, we analyze the three research questions: (RQ-1) Does RoBERTa learn long-tail tag prediction well? (RQ-2) Can a 12.5x smaller CLESS model achieve good long-tail prediction, and at what cost? (RQ-3) How does CLESS compare in zero to few-shot prediction and does its model size matter. We split the dataset into 80/10/10 for training, development, and test set. *Test scores or curves are reported for models that have the best development set average precision score AP_{micro} over all 1315 classes.* RoBERTa has 125 million parameters and is pretrained on 160GB of text data. CLESS has 8-10 million parameters and is pretrained on just 60MB of in-domain text data. We use a ZeroR classifier, i.e. predicting the majority label per class, to establish imbalanced guessing performance. The ZeroR AP_{micro} on this dataset is 0.002 since a maximum of 4 in 1315 classes are active per instance—i.e., which underlines the challenge of the task.

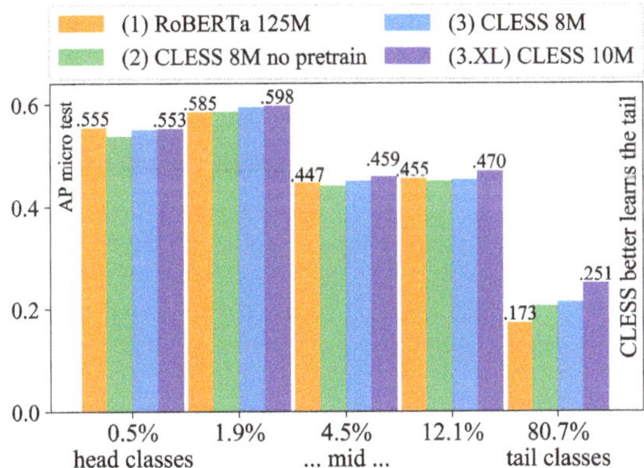

Figure 3. Long-tail performance (RQ-1, RQ-2), over all five head to tail class bins—see Figure 2. The tail class bin contains 80.7% or 1062/1315 of classes. The non-pretrained CLESS (2) underperforms, while RoBERTa performs the worst on the 80.7% of tail classes. The largest pretrained CLESS model (3.XL) outperforms RoBERTa in tail and mid class prediction, while performing nearly on par for the 7/1315 = 0.5% (most common) head classes.

5.1. (RQ-1+2): Long-Tail Capture of RoBERTa vs. CLESS

Here we compare the long-tail prediction performance of RoBERTa (1) vs. CLESS setups that, either were pretrained (3, 3.XL), or not pretrained (2). Plotting individual scores for 1315 classes is unreadable. Instead, we sort classes from frequent to rare and assign them to one of five '20% of the overall class frequency' bins, such that all bins are balanced. This means all bins contain the same amount of positive real labels (label occurrences) and are directly comparable. As seen in Figure 2, this means that the head bin (left) contains the most frequent $7/1315 = 0.5\%$ classes, while the tail contains the most rarely occurring $1061/1315 = 80.7\%$ classes.

5.1.1. RoBERTa: A Large Pretrained Model Does not Guarantee Long-Tail Capture

Figure 3 shows how a tag prediction fine-tuned RoBERTa performs over the five class bins as described above or in Section 4. RoBERTa learns the most common (0.5% head) classes well, but struggles with mid to tail classes. On the tail class bin, i.e., on $1061/1315 = 80.7\%$ of classes, RoBERTa performs worse than a CLESS model that did not use contrastive pretraining (2). This allows multiple insights. One, *a large PLM should not implicitly be assumed to learn long-tail information well.* Two, large-scale pretraining data should not be expected to contain enough (rare) long-tailed domain information for an arbitrary end-task, since in the tail-domain, data is always limited. Three, even a small supervised contrastive model, without pretraining, can improve long-tail retention (for 80.7% of classes). Together these results indicate that compressing a large PLM may not be the optimal approach to training a small, long-tail prediction capable model.

5.1.2. CLESS: Contrastive Pretraining Removes the Need for Model Compression

Model (3) and (3.XL) use our contrastive pretraining on the end-tasks' 60MB of unlabeled text data before supervised fine-tuning. We see that models with contrastive pretraining (3, 3.XL) noticeably outperform RoBERTa (1) and the non-pretrained contrastive model (2), on all non-head class bins, but especially on the 80.7% tail classes. We also see that the pretraining model parameter amount impacts CLESS performance as the 10 million parameter model (3.XL) outperforms the 8M parameters model (3) over all class bins and especially the tail bin. The above observations are especially encouraging as they tell us that contrastive in-domain pretraining can produce small, long-tail learning capable models without the need for compressing large models. It also tells us that model capacity matters in long-tail information retention, but not in the common sense that large PLMs are as useful as they have proven to be for non-long-tail learning applications. This also means that contrastive self-supervised LM pretraining can help reduce algorithmic bias caused by long-tail information loss in smaller models, the potential fairness impact of which was described by [1,2,6].

5.1.3. Practical Computational Efficiency of Contrastive Language Modeling

Though the long-tail performance results of CLESS are encouraging, its computational burden should ideally be equal or less than that of fine-tuning RoBERTa. When we analyzed training times we found that RoBERTa took 126 GPU hours to fine-tune for 48 epochs, when using 100% of fine-tuning labels. For the same task we found that CLESS (3.XL) took 7 GPU hours for self-supervised pretraining (without labels) and 5 GPU hours for supervised fine-tuning over 51 epochs—To bring CLESS to the same GPU compute load as RoBERTa ($\approx 96\%$) we parallelized our data generation—otherwise our training times double and the GPU load is only $\approx 45\%$. As a result, pretraining plus fine-tuning takes CLESS (3.XL) 12 h compared to 126 for fine-tuning RoBERTa. This means that the proposed contrastive in-domain pretraining has both qualitative and computational advantages, while remaining applicable in scenarios where large collections of pretraining data are not available—which may benefit use cases like non-English or medical NLP. Additionally, both methods benefit from parameter search, but since CLESS unifies self-supervised pretraining and supervised fine-tuning as one objective we can reuse pretraining hyperparameters during fine-tuning.

A more in-depth account of computational trade-offs is given in Appendix B, while details of hyperparameter tuning are given in Appendix C.

It is of course possible to attempt to improve the long-tail performance of RoBERTa, e.g. via continued pretraining on the in-domain data [42] or by adding new tokens [43,44]. However, this further increases the computation and memory requirements of RoBERTa, while the model still has to be compressed—which requires even more computation. We also tried to further improve the embedding initialization of CLESS using the method described in [45], to further boost its learning speed. While this helped learning very small models (<2M parameters), it did not meaningfully impact the performance of contrastive pretraining or fine-tuning.

5.2. (RQ-3.1-2): Contrastive Zero-Shot Long-Tail Learning

Thanks to the unified learning objective for self-supervised and supervised learning, CLESS enables zero-shot prediction of long-tail labels after self-supervised pretraining, i.e. without prior training on any labels. Therefore, in this section, we analyze the impact of using more model parameters (RQ-3.1) as well as using more pseudo labels (RQ-3.2) during self-supervised contrastive pretraining.

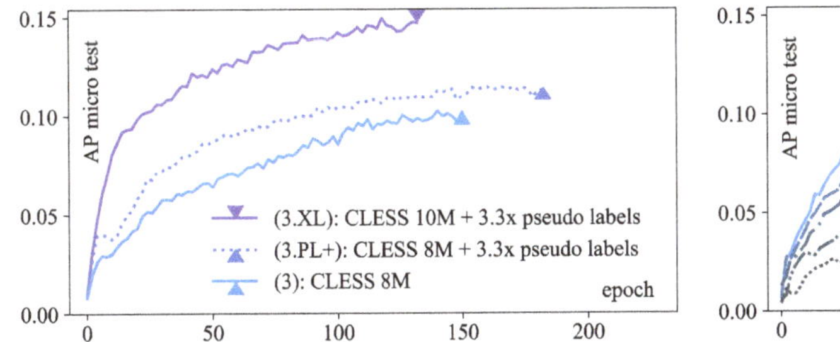

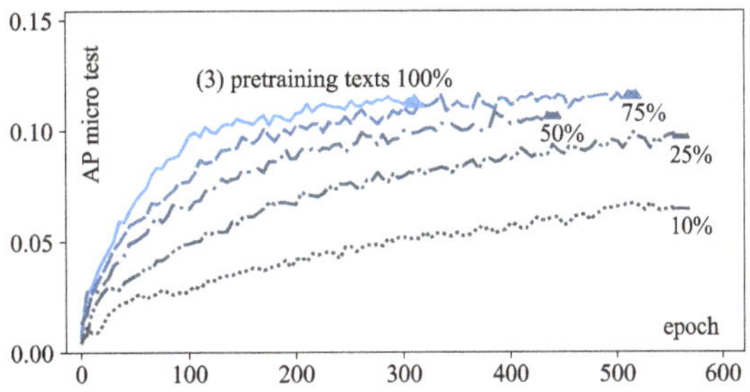

Figure 4. Zero-shot pretraining data-efficiency: by model size, pseudo label amount and pretraining text amount. Left: The zero-shot (data-efficiency) performance of the self-supervised pretraining base model (3) is increased when, adding more self-supervision pseudo labels (3.PL+) and when increasing model parameters (3.XL). Right: When only using only a proportion of the pretraining input data texts to pretrain model (3), its zero-shot learning is slowed down proportionally, but *still converges towards the 100% for all but the most extreme pretraining data reductions.*

5.2.1. (RQ-3.1): More Self-supervision and Model Size Improve Zero-Shot Long-Tail Capture

In Section 4, we study how CLESSs' zero-shot long-tail retention ability is impacted by: (left) using more pseudo labels (learning signal) during pretraining; and (right) by using only portions of unlabeled text data for pretraining. To do so, we pretrain CLESS variants on pseudo labels and evaluate each variant's zero-shot AP_{micro} performance over all 1315 classes of the real-label test set from Section 4. As before, we show test score curves for the models with the best AP_{micro} dev set performance.

The left plot of Figure 4, shows the effect of increasing the number of self-supervision pseudo label and model parameters. The CLESS 8M model (3), pretrained with 8 million parameters and 150 pseudo labels, achieves around $.10AP_{micro}$ on the test real labels as zero-shot long-tail performance. When increasing the pseudo label number to 500 in model (3.PL+), the model gains zero-shot performance (middle curve), without requiring more parameters. When additionally increasing the model parameters to 10M in (3.XL), the zero-shot performance increases substantially (top curve). Thus, both increasing self-supervision signal amount and model size boost zero-shot performance.

5.2.2. RQ-3.2: Contrastive pretraining Leads to Data-Efficient Zero-Shot Long-Tail Learning

Further, in the *right plot* of Figure 4 we see the CLESS 8M model (3) when trained on increasingly smaller portions (100%, . . . , 10%) of pretraining text. For all *but the smallest pretraining data portions (<25%)* the model still converges towards the original 100% performance. However, as expected, its convergence slows proportionally with smaller pretraining text portions since each data reduction implies seeing less pseudo label self-supervision per epoch. As a result, the data reduced setups need more training epochs, so we allowed 5*x* more waiting-epochs for early stopping than in the *left* plot. Thus, our contrastive self-supervised objective can pretrain data-effectively from very limited data. Similar data-efficiency gains from using contrastive objectives were previously only observed in computer vision applications by Zimmermann et al. [18], which confirms our initial intuition that contrastive self-supervision is generally useful for self-supervised learning from limited data.

Methods like Pappas and Henderson [15], Jiang et al. [36], Augenstein et al. [46] required supervised pretraining on real labels to later predict other, unseen labels in a zero-shot fashion. CLESS instead uses self-supervised pretraining to enable zero-shot prediction without training on real labels. This 'text-to-text' prediction approach is intentionally reminiscent of zero-shot prediction approaches in large PLMs like GPT-3 [47], but is instead designed to maximize zero-shot, long-tail prediction for use cases that strongly limit pretraining data amounts and model size. Hooker et al. [6] hypothesized that long-tail prediction depends on the model capacity (parameter amount). Additionally, Brown et al. [47] found that zero-shot prediction performance depends on model capacity, but [48,49] experimentally showed or visualized how inefficiently model capacity is used by common models, especially after fine-tuning. From the above observations, we can confirm the impact of model size for the doubly challenging task of *long-tail, zero-shot prediction*, but we can also confirm that contrastive pretraining allows a model to much more efficiently use its capacity for long-tail capture, i.e., requiring 12.5x fewer parameters (capacity) than common the RoBERTa model. Perhaps more encouragingly, we also observed that cheap, contrastive in-domain pretraining boosts zero-shot prediction, even when pretraining data is very limited—i.e. either by lack of large domain text data or due to data limitations caused by a long-tail distribution.

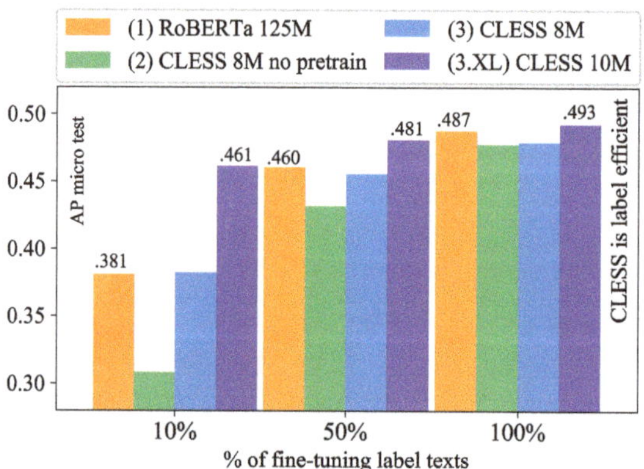

Figure 5. (RQ-3.3) Few-shot label-efficiency: (1) RoBERTa. (2) CLESS without pretraining. (3) CLESS with pretraining. (3.XL) CLESS pretrained with more pseudo labels and model parameter as described in (Section 5.2). AP_{micro_test} scores for few-shot portions: 100%, 50%, 10% of training samples with real labels. CLESS 10M outperforms RoBERTa, and retrains 93.5% of its long-tail performance using only 10% of fine-tuning label texts.

5.3. (RQ-3.3): Few-Shot Long-Tail Learning

Since CLESS models allow direct transfer (reuse) of the pretrained prediction head for supervised label prediction one would also expect the models' few-shot long-tail prediction performance to benefit from self-supervised pretraining. We thus study the few-shot learning performances of both CLESS and RoBERTa, to understand differences in large pretrained language models (PLMs) and small contrastive language model (CLM) pretraining in more detail. For the few-shot setup, we use 100%, 50% and 10% of labeled text instances for supervised training or fine-tuning of all models. This implies that if labels were common in the 100% setup, they now become increasingly rare or few-shot in the 10% setup, since the smaller label sets are still long-tail distributed. We again use AP_{micro} test set performance over all 1315 classes to compare models.

In Figure 5, we see that when using full supervision (100%), all models perform similarly, with CLESS (3.XL) slightly outperforming RoBERTa (0.493 vs. 0.487) AP_{micro_test}. For few-shot learning (10%, 50%), we see that CLESS 3.XL retrains $0.461/0.493 = 0.935\%$ of its original performance when using only 10% of fine-tuning labels, while RoBERTa and CLESS 8M each retain around 77%. This demonstrates that even a sightly larger contrastive pretraining model, with increased self-supervision signal (3.XL), not only improves zero-shot learning performance as was seen in Figure 4, but also markedly boosts few-shot performance. Noticeably, the only non-pretrained model (2), performs much worse than the others in the more restricted few-shot scenarios. Since models (2) and (3) use the same hyperparameters and only differ in being pretrained (3) or not being pretrained (2), this demonstrates that contrastive self-supervised pretraining largely improves label efficient learning.

6. Conclusion

We introduce CLESS, a contrastive self-supervised language model (CLM), that unifies self-supervised pretraining and supervised fine-tuning into a single contrastive 'text embedding to text embedding' matching objective. Through three research questions (RQ-1 to RQ-3) we demonstrate that this model learns superior zero-shot, few-shot, and fully supervised long-tail retention in *small models without needing to compress large models*. In RQ-1, we first show that a fine-tuned, large pretrained language model like RoBERTa should not implicitly be expected to learn long-tail information well. Then, in RQ-2, we demonstrate that our contrastive self-supervised pretraining objective enables very text data-efficient pretraining, which also results in markedly improved (label efficient) few-shot or zero-shot long-tail learning. Finally, in RQ-3, we find that using more contrastive self-supervision signals and increasing model parameter capacity play important roles in boosting zero to few-shot long-tail prediction performance when learning from very limited in-domain pretraining data. We also find that the very low compute requirements of our method make it a viable alternative to large pretrained language models, especially for learning from limited data or in long-tail learning scenarios, where tail data is naturally limited. In future work, we envision applying CLESS to low-data domains like medicine [27] and fact-checking [50], or to tasks where new labels emerge at test time, e.g. hashtag prediction [51]. Code and setup can be found at https://github.com/NilsRethmeier/CLESS (accessed on 30 August 2021).

Author Contributions: Conceptualization, N.R., I.A.; methodology, N.R.; software, N.R.; validation, N.R., I.A.; formal analysis, N.R., I.A.; investigation, N.R.; resources, N.R.; data curation, N.R.; writing—original draft preparation, N.R., I.A.; writing—review and editing, N.R., I.A.; visualization, N.R.; supervision, I.A.; project administration, N.R., I.A.; funding acquisition, N.R., I.A. All authors have read and agreed to the published version of the manuscript.

Funding: This research was conducted within the Cora4NLP project, which is funded by the German Federal Ministry of Education and Research (BMBF) under funding code 01IW20010.

Institutional Review Board Statement: Not applicable.

Informed Consent Statement: Not applicable.

Data Availability Statement: The dataset used in this study can be found as described in the data section. The exact training data setup is available online and upon request to the corresponding author.

Acknowledgments: We thank Yanai Elazar, Pepa Atanasova, Christoph Alt, Leonard Hennig and Dustin Wright for providing feedback and discussion on earlier drafts of this work.

Conflicts of Interest: The authors declare no conflict of interest.

Appendix A

Text preprocessing details: We decompose tags such as 'p-value' as 'p' and 'value' and split latex equations into command words, as they would otherwise create many long, unique tokens. In the future, character encodings may be better for this specific dataset, but that is out of our current research scope. Words embedding are pretrained via fastText on the training corpus text. 10 tag words are not in the input vocabulary and thus we randomly initialize their embeddings. Though we never explicitly used this information, we parsed the text and title and annotated them with 'Html-like' title, paragraph, and sentence delimiters, i.e. *</title>, </p>, and </s>*.

Appendix B

Here we will discuss the time and transfer complexity of CLESS vs. Self-attention models. We do so since time complexity is only meaningful if the data-efficiency of two methods is the same, because the combination of convergence speed, computation speed, and end-task performance makes a model effective and efficient.

Table A1. Time complexity O(Layer), data-efficiency, number of trainable parameters, number of all parameters. The data-efficiency of Convolutions (*) is reported in various works to be superior to that of self-attention models [28,30,52–56]. d is the input embedding size and its increase slows down convolutions. n is the input sequence length and slows down self-attention the most [57]. There exist optimizations for both problems.

Layer Type	$O(Layer)$	Literature Reported Data Requirements	Trainable Parameters
Convolution	$O(n \cdot d^2)$	small (*)	8M-10M (CLESS)
Self-Attention	$O(n^2 \cdot d)$	large to web-scale (*)	125M (RoBERTa)

Time complexity: Our text encoder uses a single 1D CNN encoder layer which has a complexity of $O(n \cdot k \cdot d \cdot f)$ vs. $O(n^2 \cdot d)$ for vanilla self-attention as outlined in Vaswani et al. [57]. Here n is the input sequence length, k is the convolution filter size, d is the input embedding dimension [$d = 512$ in [57] vs. $d = 100$ for us], and f is the number of convolution filters (at maximum $f = 3 \cdot 100$ for our (3.XL) pretraining model). Since we use kernel sizes $\{1,2,3\}$ we get for the largest configuration (3.XL) an $O(n \cdot k = 6 \cdot d = 1 \cdot f = 3d) \approx O(n \cdot 3d^2)$ vs. $O(n^2 \cdot 5d)$ in a vanilla (2017) self-attention setup where d = 512. Furthermore Transformer self-attention runs an n-way soft-max computation at every layer (e.g. 16 layers), while we run $g \cdot b$ single-class predictions at the final output layer using a noise contrastive objective NCE. We use NCE to undersample both: true negative learning labels (label=0) as well as positive and negative pseudo labels (input words). If the goal is to learn a specific supervised end-task, more informed sampling of positive and negative pseudo labels can be devised. However, we did not intend to overfit the supervised task by adding such hand-crafted human biases. Instead we use random sampling to pretrain a model for arbitrary downstream tasks (generalization), which follows a similar logic as random masking does in masked language modeling.

Transfer complexity: Traditional transfer NLP approaches like RoBERTa [13] need to initialize a new classification head per task which requires either training a new model per task or a joint multi-task learning setup. CLESS however can train multiple tasks, even if they arrive sequentially over time, while reusing the same classifier head from prior pretraining or fine-tuning. Thus, there is no need to retrain a separate model each time as in current Transformer transfer models. Once pretrained a CLESS model can zero-shot transfer to any new task since the match classifier is reused.

Appendix C

In this section, we describe the data and memory efficiency of the proposed method as well as the hyperparameter tuning we conducted.

Data, sample and memory efficiency: We analyzed input data and label efficiency in the main documents zero and few-shot learning sections. Regarding data-efficiency and model design choices we were guided by the existing research and optimized for data-efficient learning with inherent self-supervised zero-shot capabilities in order to facilitate and study supervision-free generalization to unforeseen tasks. We explain the origins of these design choices in more detail below. As mentioned in the related research section, Transformers rely on large to Web-scale pretraining data collections 'end-task external pre-training data' [52,53], which results in extensive pretraining hardware resources [58,59], concerns about environmental costs [56,60] and unintended contra-minority biases [56,61,62]. CNNs have been found to be more data-efficient than Transformers, i.e., train to better performance with less data, several works. For example in OPENAI's CLIP model, see Figure 2 in [30], the authors find that replacing a Transformer language model backbone with a CNN backbone increased the zero-shot data-efficiency 3 fold, which they further increased by adding a *supervised* contrastive learning objective. Ref. [38] showed that adding a CNN component to a vision Transformer model helps with data and computational efficiency, see Figure 5 and text in [38]. When comparing works on small-scale data pretraining capabilities between [54] (CNN, LSTM) with recent Transformer models Wang et al. [55], one can see that Transformer encoders struggle to learn from small pretraining collections. They also struggle to fine-tuning on smaller supervised collections [12,32,59]. For CLESS, tuning the embedding layer made little difference to end-task performance, when starting training with pretrained fastText word embedding. Thus embedding tuning the embedding layer can be turned off to reduce gradient computation and memory. For example, when not tuning embeddings, the CLESS 10M model has only 3.2M trainable parameters.

Parameter tuning + optima (2)-(3.XL) We provide detailed parameter configurations as python dictionaries for reproducibility in the code repository within the `/confs` folder. In Table A2 we see how the hyperparameters explored in CLESS—the optimal CLESS 3.XL parameters are marked in bold. The baseline CLESS configuration (2) hyperparameters were found as explained in the table, using the non-pretraining CLESS 8M (2) model—its best parameters are *italic*. We found these models by exploring hyperparameters that have been demonstrated to increase generalization and performance in [63,64]. To find optimal hyperparameter configurations for the baseline model (2) we ran a random grid search over the hyperparameter values seen in Table A2. For the baseline CLESS 8M model (2), without pretraining, we found optimal hyperparameters to be: $lr = 0.001$ (lr=0.0005 works too), $filter\_sizes\_and\_number = \{1 : 100, 2 : 100, 3 : 100\}$, $match\_classifier$=two_layer_classifier, 'conf':[{'do': None | .2, 'out_dim': 2048 | 4196 | 1024}, $max\_k_pooling$=7, bs=1536, etc.—see Table A2. Increasing the filter size, classifier size, its depth, or using larger k in k-max pooling decreased dev set performance of the non-pretrained model (i.e., CLESS 8M) due to increased overfitting. The largest pretrained CLESS 10M (3.XL) model was able to use more: 'max-k=10', a larger 'label' and 'text sequence encoder'= one_layer_label_enc, 'conf':[{'do': .2, 'out_dim': 300} while the batch size shrinks to 1024 due to increased memory requirements of label matching. Note that label and text encoder have the same output dimension in all settings—so text and label embeddings remain in the same representation dimensionality $\mathbb{R}^{300}$. The label encoder averages word embeddings (average pooling), while the text encoder uses a CNN with filters as in Table A2. The model receives text word ids and label-word ids, that are fed to the 'text encoder' and 'label-encoder'. These encoders are sub-networks that are configured via dictionaries to have fully connected layers and dropout, with optimal configurations seen in the table. As the match-classifier, which learns to contrast the (text embedding, label embedding) pairs, we use a $two\_layer_M LP$ which learns a similarity (match) function between text embedding to label embedding combinations (concatenations).

Table A2. Explored parameters. We conducted a random grid search over the following hyperparameters while optimizing important parameters first to largely limit trials. We also pre-fit the filter size, lr, and filters on a 5k training subset of samples to further reduce trails. Then, to further reduce the number of trials, we tuned in the following order: learning rate lr, filter sizes f, max-k pooling, tuning embeddings, batch size bs, and finally the depth of the matching-classifier MLP. This gave us a baseline model, (2) CLESS 8M, that does not use pretraining to save trials and compute costs, but could be used to build up into the self-supervised pretraining models (3) and (3.XL) by increasing self-supervision and model size. Fortunately, RoBERTa has established default parameters reported in both its code documentation (https://github.com/pytorch/fairseq/tree/master/examples/roberta) (accessed on 30 September 2021) and the https://simpletransformers.ai (accessed on 30 September 2021) version, where we varied batch size, warmup, and learning rate around the default setting of these sources. Below we give the search parameters for CLESS. For CLESS 8M (2,3) the best params are *italic* and for CLESS 10M (3.XL) the best params are **bold**.

Filter size: num filters	{1: 57, 2: 29, 3: 14}, {**1**:100, **2**:100, **1**:100},{1: 285, 2: 145, 3: 70}, {1:10, 10:10, 1:10}, {1:15, 2:10, 3:5}, {1:10}, {1:100}, {10:100}
lr	0.01, 0.0075, 0.005, ***0.001***, 0.0005, 0.0001
bs (match size)	**1024**, *1536*, 4096
max-k	1, 3, *7*, **10**
match-classifier	***two_layer_classifier***, 'conf':[{'do': None|**.2**, 'out_dim': **2048**|4196|*1024*}, {'do':***None***|0.2}]}, one_layer_classifier, 'conf':[{'do':.2}]}
label encoder	*one_layer_label_enc, 'conf':[{'do':* None|.2, *'out_dim': 100}*, one_layer_label_enc, **'conf':[{'do': .2, 'out_dim': 300}**
seq encoder	*one_layer_label_enc, 'conf':[{'do':* None|.2, *'out_dim': 100}*, one_layer_label_enc, **'conf':[{'do': .2, 'out_dim': 300}**
tune embedding:	True, False
#real label samples:	20, *150*, **500** (g positives (as annotated in dataset), b random negative labels—20 works well too)
#pseudo label samples:	20, 150, **500** (g positives input words, b negative input words)—used for self-superv. pretraining
optimizer:	**ADAM**—default params, except lr

Parameter tuning + optima (2)-(3.XL) We provide detailed parameter configurations as python dictionaries for reproducibility in the code repository within the `/confs` folder. In Table A2 we see how the hyperparameters explored in CLESS—the optimal CLESS 3.XL parameters are marked in bold. The baseline CLESS configuration (2) hyperparameters were found as explained in the table, using the non-pretraining CLESS 8M (2) model—its best parameters are *italic*. We found these models by exploring hyperparameters that have been demonstrated to increase generalization and performance in [63,64]. To find optimal hyperparameter configurations for the baseline model (2) we ran a random grid search over the hyperparameter values seen in Table A2. For the baseline CLESS 8M model (2), without pretraining, we found optimal hyperparameters to be: $lr = 0.001$ (lr=0.0005 works too), $filter\_sizes\_and\_number = \{1 : 100, 2 : 100, 3 : 100\}$, $match\_classifier$=two_layer_classifier, 'conf':[{'do': None | .2, 'out_dim': 2048 | 4196 | 1024}, $max\_k_pooling$=7, bs=1536, etc.—see Table A2. Increasing the filter size, classifier size, its depth, or using larger k in k-max pooling decreased dev set performance of the non-pretrained model (i.e., CLESS 8M) due to increased overfitting. The largest pretrained CLESS 10M (3.XL) model was able to use more: 'max-k=10', a larger 'label' and 'text sequence encoder'= one_layer_label_enc, 'conf':[{'do': .2, 'out_dim': 300} while the batch size shrinks to 1024 due to increased memory requirements of label matching. Note that label and text encoder have the same output dimension in all settings—so text and label embeddings remain in the same representation dimensionality $\mathbb{R}^{300}$. The label encoder averages word embeddings (average pooling), while the text encoder uses a CNN with filters as in Table A2. The model receives text word ids and label-word ids, that are fed to the 'text encoder' and 'label-encoder'. These encoders are sub-networks that are configured via dictionaries to have fully connected layers and dropout, with optimal configurations seen in the table. As the match-classifier, which learns to contrast the (text embedding, label embedding) pairs, we use a *two_layer$_M$LP* which learns a similarity (match) function between text embedding to label embedding combinations (concatenations).

During self-supervised pretraining, the models (3) and (3.XL) optimize for *arbitrary unforeseen long-tail end-tasks*, which allows zero-shot prediction without ever seeing real labels, but also uses a very diverse learning signal by predicting sampled positive and negative input word embeddings. If the goal is to solely optimize for a specific end-task, this self-supervision signal can be optimized to pretrain much faster, e.g. by only sampling specific word types like nouns or named entities. With specific end-task semantics in mind, the pseudo label and input manipulations can easily be adjusted. This allows adding new self-supervision signals without a need to touch the model's network code directly, which helps ease application to new tasks and for less experienced machine learning practitioners. Finally, we mention implementation features, that can safely be avoided to reduce computation and optimization effort, so that following research needs not explore this option. When training the supervised and self-supervised loss at the same time (jointly), CLESS rescales both batch losses to be of the same loss value as using a single loss. This makes it easy to balance (weight) the two loss contributions in learning, and allows transferring hyperparameters between self-supervised and supervised pretraining. We also allow re-weighting the loss balance by a percentage, so that one loss can dominate. However, we found that in practice: (a) using the self-supervised loss along with the supervised one does not improve quality, but slows computation (2 losses). (b) We also found that, if one decides to use joint self and supervised training, loss re-weighting had no marked quality effects, and should be left at 1.0 (equal weighting), especially since it otherwise introduces further, unnecessary hyperparameters. For pretraining research, hyperparameter search is very involved, because we deviate in common practice by introducing a new architecture, a new loss variation, an uncommon optimization goal and metrics as well as a new dataset. Thus we ended up with 205 trails for small test set, RoBERTa, CLESS variants, zero-shot and few shot hyperparameter search. On the herein reported dataset, we have not yet tested further scaling up model parameters for pretraining as this goes against the goal of the paper and is instead investigated in followup work. Furthermore, when we ran such parameter scale-up experiments, to guarantee empirical insights, these created a significant portion of trails, meaning that, now that sensible parameters are established, we can use much fewer trials, as is the case with pretrained transformers. The work at hand suggest that, once sensible parameters are established, they are quite robust, such that doubling the learning rate, batch size and loss weighting only cause moderate performance fluctuations. Finally, the above reported pretraining hyperparameters seem to work well on currently developed followup research, that uses other, even much larger, datasets. This makes the 205 hyperparameter trials a one time investment for initial pretraining hyperparameter (re)search for this contrastive language model (CLESS).

References

1. Hooker, S.; Courville, A.; Clark, G.; Dauphin, Y.; Frome, A. What Do Compressed Deep Neural Networks Forget? *arXiv* **2020**, arXiv:1911.05248.
2. Hooker, S. Moving beyond "algorithmic bias is a data problem". *Patterns* **2021**, 2, 100241. [CrossRef] [PubMed]
3. Liu, Z.; Miao, Z.; Zhan, X.; Wang, J.; Gong, B.; Yu, S.X. Large-Scale Long-Tailed Recognition in an Open World. In Proceedings of the IEEE CVPR, Long Beach, CA, USA, 16–20 June 2019; pp. 2537–2546. [CrossRef]
4. D'souza, D.; Nussbaum, Z.; Agarwal, C.; Hooker, S. A Tale of Two Long Tails. *arXiv* **2021**, arXiv:2107.13098.
5. Hu, N.T.; Hu, X.; Liu, R.; Hooker, S.; Yosinski, J. When does loss-based prioritization fail? *arXiv* **2021**, arXiv:2107.07741.
6. Hooker, S.; Moorosi, N.; Clark, G.; Bengio, S.; Denton, E. Characterizing and Mitigating Bias in Compact Models. In Proceedings of the 5th ICML Workshop on Human Interpretability in Machine Learning (WHI), Virtual Conference, 17 July 2020.

7. Zhuang, D.; Zhang, X.; Song, S.L.; Hooker, S. Randomness In Neural Network Training: Characterizing The Impact of Tooling. *arXiv* **2021**, arXiv:2106.11872.
8. Chang, W.C.; Yu, H.F.; Zhong, K.; Yang, Y.; Dhillon, I. X-BERT: eXtreme Multi-label Text Classification with using Bidirectional Encoder Representations from Transformers. *arXiv* **2019**, arXiv:1905.02331.
9. Joseph, V.; Siddiqui, S.A.; Bhaskara, A.; Gopalakrishnan, G.; Muralidharan, S.; Garland, M.; Ahmed, S.; Dengel, A. Reliable model compression via label-preservation-aware loss functions. *arXiv* **2020**, arXiv:2012.01604.
10. Blakeney, C.; Huish, N.; Yan, Y.; Zong, Z. Simon Says: Evaluating and Mitigating Bias in Pruned Neural Networks with Knowledge Distillation. *arXiv* **2021**, arXiv:2106.07849.
11. Jiang, Z.; Chen, T.; Mortazavi, B.J.; Wang, Z. Self-Damaging Contrastive Learning. *Proc. Mach. Learn. Res. PMLR* **2021**, *139*, 4927–4939.
12. Rogers, A.; Kovaleva, O.; Rumshisky, A. A Primer in BERTology: What We Know About How BERT Works. *Trans. Assoc. Comput. Linguist.* **2020**, *8*, 842–866. [CrossRef]
13. Liu, Y.; Ott, M.; Goyal, N.; Du, J.; Joshi, M.; Chen, D.; Levy, O.; Lewis, M.; Zettlemoyer, L.; Stoyanov, V. RoBERTa: A Robustly Optimized BERT Pretraining Approach. *arXiv* **2019**, arXiv:1907.11692.
14. Liu, J.; Chang, W.; Wu, Y.; Yang, Y. Deep Learning for Extreme Multi-label Text Classification. In Proceedings of the 40th International ACM SIGIR Conference on Research and Development in Information Retrieval, Shinjuku, Tokyo, Japan, 7–11 August 2017; pp. 115–124. [CrossRef]
15. Pappas, N.; Henderson, J. GILE: A Generalized Input-Label Embedding for Text Classification. *Trans. Assoc. Comput. Linguistics* **2019**, *7*, 139–155. [CrossRef]
16. Raffel, C.; Shazeer, N.; Roberts, A.; Lee, K.; Narang, S.; Matena, M.; Zhou, Y.; Li, W.; Liu, P.J. Exploring the Limits of Transfer Learning with a Unified Text-to-Text Transformer. *J. Mach. Learn. Res.* **2020**, *21*, 1–67.
17. Graf, F.; Hofer, C.; Niethammer, M.; Kwitt, R. Dissecting Supervised Constrastive Learning. *Proc. Mach. Learn. Res. PMLR* **2021**, *139*, 3821–3830.
18. Zimmermann, R.S.; Sharma, Y.; Schneider, S.; Bethge, M.; Brendel, W. Contrastive Learning Inverts the Data Generating Process. *Proc. Mach. Learn. Res. PMLR* **2021**, *139*, 12979–12990.
19. Zhang, H.; Xiao, L.; Chen, W.; Wang, Y.; Jin, Y. Multi-Task Label Embedding for Text Classification. In Proceedings of the 2018 Conference on Empirical Methods in Natural Language Processing, Brussels, Belgium, 2–4 November 2018; Association for Computational Linguistics: Stroudsburg, PA, USA, 2018; pp. 4545–4553. [CrossRef]
20. Musgrave, K.; Belongie, S.J.; Lim, S. A Metric Learning Reality Check. In Proceedings of the Computer Vision-ECCV 2020-16th European Conference, Glasgow, UK, 23–28 August 2020; pp. 681–699. [CrossRef]
21. Rethmeier, N.; Augenstein, I. A Primer on Contrastive Pretraining in Language Processing: Methods, Lessons Learned and Perspectives. *arXiv* **2021**, arXiv:2102.12982.
22. Saunshi, N.; Plevrakis, O.; Arora, S.; Khodak, M.; Khandeparkar, H. A Theoretical Analysis of Contrastive Unsupervised Representation Learning. *Proc. Mach. Learn. Res. PMLR* **2019**, *97*, 5628–5637.
23. Khosla, P.; Teterwak, P.; Wang, C.; Sarna, A.; Tian, Y.; Isola, P.; Maschinot, A.; Liu, C.; Krishnan, D. Supervised Contrastive Learning. In *Advances in Neural Information Processing Systems 33 (NeurIPS 2020)*; Larochelle, H., Ranzato, M., Hadsell, R., Balcan, M.F., Lin, H., Eds.; Curran Associates, Inc.: Red Hook, NY, USA, 2020; Volume 33, pp. 18661–18673.
24. Ostendorff, M.; Rethmeier, N.; Augenstein, I.; Gipp, B.; Rehm, G. Neighborhood Contrastive Learning for Scientific Document Representations with Citation Embeddings. *arXiv* **2022**, arXiv:2202.06671.
25. Wang, T.; Isola, P. Understanding Contrastive Representation Learning through Alignment and Uniformity on the Hypersphere. *Proc. Mach. Learn. Res. PMLR* **2020**, *119*, 9929–9939.
26. Mnih, A.; Teh, Y.W. A Fast and Simple Algorithm for Training Neural Probabilistic Language Models. In Proceedings of the 29th International Conference on International Conference on Machine Learning (ICML'12), Edinburgh, UK, 26 June–1 July 2012; Omnipress: Madison, WI, USA; pp. 419–426.
27. Şerbetci, O.N.; Möller, S.; Roller, R.; Rethmeier, N. EffiCare: Better Prognostic Models via Resource-Efficient Health Embeddings. *AMIA Annu. Symp. Proc.* **2020**, *2020*, 1060–1069. PMCID: PMC8075498.
28. Kim, K.M.; Hyeon, B.; Kim, Y.; Park, J.H.; Lee, S. Multi-pretraining for Large-scale Text Classification. In *Findings of the Association for Computational Linguistics: EMNLP 2020*, Cohn, T., He, Y., Liu Y., Eds.; Association for Computational Linguistics: Stroudsburg, PA, USA, 2020; pp. 2041–2050. [CrossRef]
29. Tay, Y.; Dehghani, M.; Gupta, J.; Bahri, D.; Aribandi, V.; Qin, Z.; Metzler, D. Are Pre-trained Convolutions Better than Pre-trained Transformers? *arXiv* **2021**, arXiv:2105.03322.
30. Radford, A.; Kim, J.W.; Hallacy, C.; Ramesh, A.; Goh, G.; Agarwal, S.; Sastry, G.; Askell, A.; Mishkin, P.; Clark, J.; et al. Learning Transferable Visual Models From Natural Language Supervision. *Proc. Mach. Learn. Res. PMLR* **2021**, *139*, 8748–8763.
31. Devlin, J.; Chang, M.; Lee, K.; Toutanova, K. BERT: Pre-training of Deep Bidirectional Transformers for Language Understanding. In Proceedings of the NAACL-HLT 2019, Minneapolis, MN, USA, 2–7 June 2019; Association for Computational Linguistics, Stroudsburg, PA, USA, 2019; pp. 4171–4186. [CrossRef]
32. Liu, L.; Liu, X.; Gao, J.; Chen, W.; Han, J. Understanding the Difficulty of Training Transformers. *arXiv* **2020**, arXiv:2004.08249.

33. Szegedy, C.; Vanhoucke, V.; Ioffe, S.; Shlens, J.; Wojna, Z. Rethinking the Inception Architecture for Computer Vision. In Proceedings of the 2016 IEEE Conference on Computer Vision and Pattern Recognition, CVPR 2016, Las Vegas, NV, USA, 27–30 June 2016. pp. 2818–2826. [CrossRef]
34. Gao, T.; Yao, X.; Chen, D. SimCSE: Simple Contrastive Learning of Sentence Embeddings. *arXiv* **2021**, arXiv:2104.08821.
35. Ma, Z.; Collins, M. Noise Contrastive Estimation and Negative Sampling for Conditional Models: Consistency and Statistical Efficiency. In Proceedings of the 2018 Conference on Empirical Methods in Natural Language Processing, Brussels, Belgium, 31 October–4 November 2018; pp. 3698–3707. [CrossRef]
36. Jiang, H.; Wang, R.; Shan, S.; Chen, X. Transferable Contrastive Network for Generalized Zero-Shot Learning. In Proceedings of the 2019 IEEE/CVF International Conference on Computer Vision, ICCV 2019, Seoul, Korea, 27 October–2 November 2019; pp. 9764–9773. [CrossRef]
37. Simoncelli, E.P.; Olshausen, B.A. Natural image statistics and neural representation. *Annu. Rev. Neurosci.* **2001**, *24*, 1193–1216. [CrossRef] [PubMed]
38. Dosovitskiy, A.; Beyer, L.; Kolesnikov, A.; Weissenborn, D.; Zhai, X.; Unterthiner, T.; Dehghani, M.; Minderer, M.; Heigold, G.; Gelly, S.; et al. An Image is Worth 16x16 Words: Transformers for Image Recognition at Scale. *arXiv* **2020**, arXiv:2010.11929.
39. Lukasik, M.; Bhojanapalli, S.; Menon, A.; Kumar, S. Does label smoothing mitigate label noise? *Proc. Mach. Learn. Res. PMLR* **2020**, *119*, 6448–6458.
40. Davis, J.; Goadrich, M. The relationship between Precision-Recall and ROC curves. In Proceedings of the 23rd International Conference on Machine Learning, (ICML 2006), Pittsburgh, PA, USA, 25–29 June 2006; pp. 233–240. [CrossRef]
41. Fernández, A.; García, S.; Galar, M.; Prati, R.C.; Krawczyk, B.; Herrera, F. *Learning from Imbalanced Data Sets*; Springer: Berlin/Heidelberg, Germany, 2018. [CrossRef]
42. Gururangan, S.; Marasović, A.; Swayamdipta, S.; Lo, K.; Beltagy, I.; Downey, D.; Smith, N.A. Don't Stop Pretraining: Adapt Language Models to Domains and Tasks. In Proceedings of the 58th Annual Meeting of the Association for Computational Linguistics, Virtual Conference, 6–8 July 2020; Association for Computational Linguistics: Stroudsburg, PA, USA, 2020; pp. 8342–8360. [CrossRef]
43. Poerner, N.; Waltinger, U.; Schütze, H. Inexpensive Domain Adaptation of Pretrained Language Models: Case Studies on Biomedical NER and Covid-19 QA. In *Findings of the Association for Computational Linguistics: EMNLP 2020*, Cohn, T., He, Y., Liu Y., Eds.; Association for Computational Linguistics: Stroudsburg, PA, USA, 2020; pp. 1482–1490. [CrossRef]
44. Tai, W.; Kung, H.T.; Dong, X.; Comiter, M.; Kuo, C.F. exBERT: Extending Pre-trained Models with Domain-specific Vocabulary Under Constrained Training Resources. In *Findings of the Association for Computational Linguistics: EMNLP 2020*, Cohn, T., He, Y., Liu Y., Eds.; Association for Computational Linguistics: Stroudsburg, PA, USA, 2020; pp. 1433–1439. [CrossRef]
45. Rethmeier, N.; Plank, B. MoRTy: Unsupervised Learning of Task-specialized Word Embeddings by Autoencoding. In Proceedings of the 4th Workshop on Representation Learning for NLP, RepL4NLP@ACL 2019, Florence, Italy, 2 August 2019; pp. 49–54. [CrossRef]
46. Augenstein, I.; Ruder, S.; Søgaard, A. Multi-Task Learning of Pairwise Sequence Classification Tasks over Disparate Label Spaces. In Proceedings of the 2018 Conference of the North American Chapter of the Association for Computational Linguistics: Human Language Technologies, Volume 1 (Long Papers), New Orleans, LA, USA, 1–6 June 2018; Association for Computational Linguistics: Stroudsburg, PA, USA, 2018; pp. 1896–1906. [CrossRef]
47. Brown, T.B.; Mann, B.; Ryder, N.; Subbiah, M.; Kaplan, J.; Dhariwal, P.; Neelakantan, A.; Shyam, P.; Sastry, G.; Askell, A.; et al. Language Models are Few-Shot Learners. *Adv. Neural Inf. Process. Syst.* **2020**, *33*, 1877–1901.
48. Frankle, J.; Carbin, M. The Lottery Ticket Hypothesis: Finding Sparse, Trainable Neural Networks. In Proceedings of the 7th International Conference on Learning Representations, ICLR 2019, New Orleans, LA, USA, 6–9 May 2019.
49. Rethmeier, N.; Saxena, V.K.; Augenstein, I. TX-Ray: Quantifying and Explaining Model-Knowledge Transfer in (Un-)Supervised NLP. In Proceedings of the Proceedings of the Thirty-Sixth Conference on Uncertainty in Artificial Intelligence, UAI 2020, Toronto, ON, Canada, 3–6 August 2020.
50. Augenstein, I.; Lioma, C.; Wang, D.; Chaves Lima, L.; Hansen, C.; Hansen, C.; Simonsen, J.G. MultiFC: A Real-World Multi-Domain Dataset for Evidence-Based Fact Checking of Claims. In Proceedings of the 2019 Conference on Empirical Methods in Natural Language Processing and the 9th International Joint Conference on Natural Language Processing (EMNLP-IJCNLP), Hong Kong, China, 3–7 November 2019; Association for Computational Linguistics: Stroudsburg, PA, USA, 2019; pp. 4685–4697. [CrossRef]
51. Ma, Z.; Sun, A.; Yuan, Q.; Cong, G. Tagging Your Tweets: A Probabilistic Modeling of Hashtag Annotation in Twitter. In Proceedings of the 23rd ACM International Conference on Information and Knowledge Management (CIKM '14), Shanghai, China, 3–7 November 2014; Li, J., Wang, X.S., Garofalakis, M.N., Soboroff, I., Suel, T., Wang, M., Eds.; Association for Computing Machinery: New York, NY, USA, 2014; pp. 999–1008.
52. Liu, Q.; Kusner, M.J.; Blunsom, P. A Survey on Contextual Embeddings. *arXiv* **2020**, arXiv:2003.07278.
53. Yogatama, D.; de Masson d'Autume, C.; Connor, J.; Kociský, T.; Chrzanowski, M.; Kong, L.; Lazaridou, A.; Ling, W.; Yu, L.; Dyer, C.; et al. Learning and Evaluating General Linguistic Intelligence. *arXiv* **2019**, arXiv:1901.11373.
54. Merity, S.; Xiong, C.; Bradbury, J.; Socher, R. Pointer Sentinel Mixture Models. *arXiv* **2017**, arXiv:1609.07843.
55. Wang, C.; Ye, Z.; Zhang, A.; Zhang, Z.; Smola, A.J. Transformer on a Diet. *arXiv* **2020**, arXiv:2002.06170.

56. Bender, E.M.; Gebru, T.; McMillan-Major, A.; Shmitchell, S. On the Dangers of Stochastic Parrots: Can Language Models Be Too Big? In Proceedings of the 2021 ACM Conference on Fairness, Accountability, and Transparency (FAccT '21), Virtual Event, 3–10 March 2021; Association for Computing Machinery: New York, NY, USA, 2021; pp. 610–623. [CrossRef]
57. Vaswani, A.; Shazeer, N.; Parmar, N.; Uszkoreit, J.; Jones, L.; Gomez, A.N.; Kaiser, L.; Polosukhin, I. Attention is All you Need. In *Advances in Neural Information Processing Systems 30 (NIPS 2017)*; Curran Associates Inc.: Red Hook, NY, USA, 2017; ISBN: 9781510860964.
58. Hooker, S. The Hardware Lottery. *Commun. ACM* **2020**, *64*, 58–65. [CrossRef]
59. Dodge, J.; Ilharco, G.; Schwartz, R.; Farhadi, A.; Hajishirzi, H.; Smith, N.A. Fine-Tuning Pretrained Language Models: Weight Initializations, Data Orders, and Early Stopping. *arXiv* **2020**, arXiv:2002.06305.
60. Strubell, E.; Ganesh, A.; McCallum, A. Energy and Policy Considerations for Deep Learning in NLP. In Proceedings of the 57th Annual Meeting of the Association for Computational Linguistics, Florence, Italy, 28 July–2 August 2019; Association for Computational Linguistics: Stroudsburg, PA, USA, 2019. pp. 3645–3650. [CrossRef]
61. Mitchell, M.; Baker, D.; Moorosi, N.; Denton, E.; Hutchinson, B.; Hanna, A.; Gebru, T.; Morgenstern, J. Diversity and Inclusion Metrics in Subset Selection. In Proceedings of the AAAI/ACM Conference on AI, Ethics, and Society, New York, NY, USA, 7–9 February 2020; Association for Computing Machinery: New York, NY, USA, 2020. pp. 117–123. [CrossRef]
62. Waseem, Z.; Lulz, S.; Bingel, J.; Augenstein, I. Disembodied Machine Learning: On the Illusion of Objectivity in NLP. *arXiv* **2021**, arXiv:2101.11974.
63. Jiang, Y.; Neyshabur, B.; Mobahi, H.; Krishnan, D.; Bengio, S. Fantastic Generalization Measures and Where to Find Them. *arXiv* **2019**, arXiv:1912.02178.
64. He, F.; Liu, T.; Tao, D. Control Batch Size and Learning Rate to Generalize Well: Theoretical and Empirical Evidence. In *NeurIPS*; Wallach, H., Larochelle, H., Beygelzimer, A., d'Alché Buc, F., Fox, E., Garnett, R., Eds.; Curran Associates, Inc.: Red Hook, NY, USA, 2019.

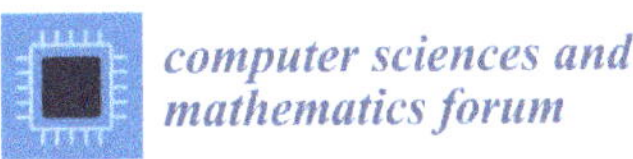

computer sciences and mathematics forum

Proceeding Paper

Age Should Not Matter: Towards More Accurate Pedestrian Detection via Self-Training †

Shunsuke Kogure [1,‡], Kai Watabe [1,‡], Ryosuke Yamada [1], Yoshimitsu Aoki [2], Akio Nakamura [3] and Hirokatsu Kataoka [2,*]

1 National Institute of Advanced Industrial Science and Research Technology (AIST), Tokyo 135-0064, Japan; skogure@aoki-medialab.jp (S.K.); kai.watabe@aist.go.jp (K.W.); ryosuke.yamada@aist.go.jp (R.Y.)
2 Department of Electronics and Electrical Engineering, Keio University, Yokohama 223-8522, Japan; aoki@elec.keio.ac.jp
3 Department of Robotics and Mechatronics, Tokyo Denki University, Tokyo 120-0026, Japan; nkmr-a@cck.dendai.ac.jp
* Correspondence: hirokatsu.kataoka@aist.go.jp
† Presented at the AAAI Workshop on Artificial Intelligence with Biased or Scarce Data (AIBSD), Online, 28 February 2022.
‡ These authors contributed equally to this work.

Abstract: Why is there disparity in the miss rates of pedestrian detection between different age attributes? In this study, we propose to (i) improve the accuracy of pedestrian detection using our pre-trained model; and (ii) explore the causes of this disparity. In order to improve detection accuracy, we extend a pedestrian detection pre-training dataset, the Weakly Supervised Pedestrian Dataset (WSPD), by means of self-training, to construct our Self-Trained Person Dataset (STPD). Moreover, we hypothesize that the cause of the miss rate is due to three biases: (1) the apparent bias towards "adults" versus "children"; (2) the quantity of training data bias against "children"; and (3) the scale bias of the bounding box. In addition, we constructed an evaluation dataset by manually annotating "adult" and "child" bounding boxes to the INRIA Person Dataset. As a result, we confirm that the miss rate was reduced by up to 0.4% for adults and up to 3.9% for children. In addition, we discuss the impact of the size and appearance of the bounding boxes on the disparity in miss rates and provide an outlook for future research.

Keywords: computer vision; pedestrian detection; fairness

Citation: Kogure, S.; Watabe, K.; Yamada, R.; Aoki, Y.; Nakamura, A.; Kataoka, H. Age Should Not Matter: Towards More Accurate Pedestrian Detection via Self-Training. *CSFM* **2022**, 3, 11. https://doi.org/10.3390/cmsf2022003011

Academic Editors: Kuan Chuan Peng and Ziyan Wu

Published: 24 May 2022

Publisher's Note: MDPI stays neutral with regard to jurisdictional claims in published maps and institutional affiliations.

1. Introduction

Recently, research has frequently explored approaches to pedestrian detection which is expected to be applied in various fields. The remarkable progress that has been made in this area is partly due to the large-scale collection of human images from the Web.

However, there are still concerns about the safety of utilizing pedestrian detection in areas such as automated driving. One of these concerns is the disparity in detection rates based on human age and race; specifically, a disparity in detection rates between "adults" and "children" has been reported when using classical human detection methods. Brando [1] affirmed that the difference in the quantity of adult versus child data in the person detection dataset is a problem that naturally arises from demographics. There are a small number of "children" in the existing pedestrian dataset, which we assume is responsible for a sample bias and a detection rate disparity between "adults" and "children."

In this paper, we constructed our Self-Trained Person Dataset (STPD) by extending the Weakly Supervised Pedestrian Dataset (WSPD) [2] to improve the accuracy of person detection. We studied the effect of each age attribute on detection performance using each pre-trained model generated by the WSPD and STPD. The INRIA Person Dataset [3] is used to evaluate the detection performance. We re-annotated both the training and test data of the INRIA Person Dataset to rigorously investigate the effect of age on the accuracy of pedestrian detection. For this re-annotation, we added the age attribute and the bounding box (bbox). In this way, we constructed a dataset for pedestrian detection validation with the age attribute. In addition, we studied the reason for the disparity in detection rate by age. Specifically, we examined the age gap in the detection rate using three experiments: (i) we clarify whether there is a difference in appearance between "adults" and "children"; (ii) we study the impact of the data augmentation of children's learning data alone on the missed rate; and (iii) finally, we compare the miss rate for each age attribute when the scale of the input image is changed. Our contributions are as follows:

- The STPD was constructed by extending the pedestrian dataset, WSPD, using self-training.
- In order to rigorously evaluate the detection performance for "adults" versus "children," we constructed a new evaluation dataset.
- The person detector with STPD pre-training reduced the miss rate of "adults" and "children" compared to the detector with WSPD pre-training. Furthermore, we observed a mitigating effect of self-training on the detection rate gap.
- We studied three aspects to investigate the cause of the gap in detection rates by age: (i) the appearance of "adults" and "children"; (ii) the quantity of data for "children"; and (iii) the scale of the input images.

2. Related Work

2.1. Detector

In recent years, approaches to detection have been dramatically improved with the rise of deep neural networks (DNNs). In the literature, a two-step region identifier and DNN-based classification have been proposed [4]. The basic approach, called R-CNN, follows three steps when generating bounding boxes: (i) detecting areas in the image that may contain objects (region proposal); (ii) extracting CNN features from region candidates; and (iii) classifying objects based on the extracted features. Fast R-CNN [5] also generates region proposals, but it is more efficient than R-CNN because Fast R-CNN pools the CNN features corresponding to each region proposal. Faster R-CNN [6] adds a region proposal network (RPN) to generate a region proposal in the network. Current research focuses on widely divided one-shot detectors such as single-shot multi-box detector (SSD) [7] and you look only once (YOLO) [8].

Recent works have also focused on high-performance detectors, such as M2Det [9], RetinaNet [10] and instance segmentation with Mask R-CNN [11]. In this paper, we applied SSD as a method of detecting people in a dataset. Here, we used a WSPD pre-trained model for self-training.

2.2. Pedestrian Detection

In the past decade, approaches to person detection have dramatically improved. Recent work has proposed configurations to improve recognition and localization, including DNNs, semantic segmentation, combined methods and small image and cloud analysis. However, in order to train these models, it is necessary to prepare a large dataset and fine-tune its architecture (e.g., SSD or M2Det). Wilson et al. tested whether an object detector can correctly detect pedestrians with different skin colors [12]. In addition, they found that it is problematic to accurately detect children because their miss rate is higher than that of adults [1]. In this study, we were able to detect pedestrians more reliably than in previous studies.

3. Self-Training

3.1. Problem

A number of datasets for pedestrian detection have been proposed to date. However, as shown in Table 1, their scale is small compared to those used for object detection. Minoguchi et al. proposed a weakly supervised learning method that eliminates false positives using existing pre-trained models by referring to bounding boxes and SVM and by constructing a labeled dataset called the Weakly Supervised Person Dataset (WSPD) [2], which far exceeds the scale of previous pedestrian detection datasets. To the best of our knowledge, the WSPD is the largest existing pedestrian dataset. Minoguchi et al. revealed the detection performance of the pre-trained model on that dataset but did not mention the disparity in the miss rate for each age attribute. Table 2 shows the attribute distribution of some bounding boxes in the WSPD. This distribution is based on our random selection of 5000 bounding boxes from the WSPD and their classification by attribute. The "Noise" label indicates that there is no person in the bounding box, while the "Multiple" label indicates that the bounding box contains multiple people. As such, we can see that the existing pedestrian dataset has a large bias in the distribution of the quantity of data; in particular, the data for children are excessively limited. Therefore, it is necessary to check whether this bias in the quantity of data contributes to the disparity in detection performance.

Table 1. Comparison of object detection and person detection datasets.

Dataset	Image	Bounding Box	Class
Pascal VOC	11,530	27,450	20
MS COCO	123,287	896,782	80
OpenImages v5	1,743,042	14,610,229	600
CityPersons	5000	35,016	2
EuroCity Persons	47,300	238,200	17
Caltech Pedestrian	250,000	350,000	2
WSPD	2,822,421	8,716,461	2
STPD (Ours)	3,461,024	9,739,996	1
FA-INRIA (Ours)	902	2993	2

Table 2. The age attribute statistics for people in bounding boxes in 5000 randomly sampled images from the WSPD dataset. The "Noise" label indicates that there is no person in the bounding box, whereas "Multiple" label means that one bounding box contains multiple people. In this paper, images labeled "Multiple" are not considered.

Annotation Type	Images	%
(i) Adult	2687	53.7
(ii) Children	169	3.4
(iii) Noise	536	10.7
(iv) Multiple	1608	32.2

3.2. Solution

As previously mentioned, we can see that the WSPD contains the largest number of images and bounding boxes among the available person detection datasets. Furthermore, the WSPD contains a wide variety of person images collected from various locations around the world. The semi-automatically collected dataset has millions of bounding boxes which may be useful for pre-training. We used a WSPD pre-trained model to apply self-training to another dataset to collect high-quality bounding boxes and to investigate the impact of each age attribute on the miss rate. Our self-training pipeline is shown in Figure 1. First, we input images from the Places365 dataset [13] to the SSD, a detector pre-trained with the WSPD, to estimate the location of the bounding box. We assigned a pseudo-label of "person" to the predicted bounding box. The determination of the location of the bounding box when generating the pseudo-label is expressed by the following equation:

$$(y', b'_{box}) = D(x;\theta), \tag{1}$$

where y' and b' represent the predicted values of the object category and bounding box, respectively, and θ represents the learned parameters of the detector. Our self-training approach allows us to automatically extend the dataset. We refer to the WSPD and the generated pseudo-labeled Places365 dataset together as the Self-Trained Person Dataset (STPD).

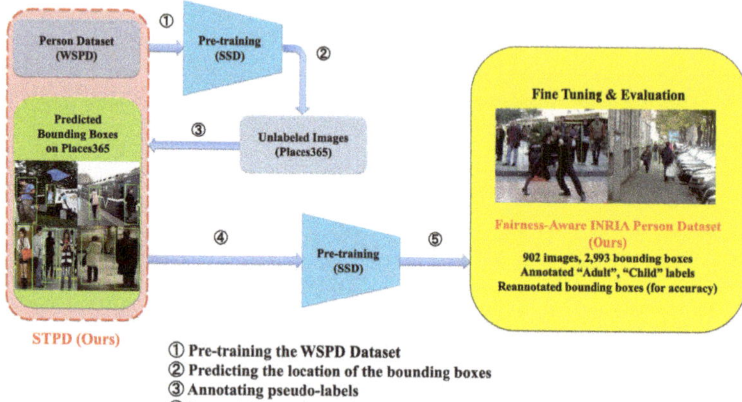

Figure 1. The self-training approach. We used a model pre-trained using the WSPD dataset with the SSD to infer the location of bounding boxes for unlabeled images from the Places365 dataset. We then gave each predicted bounding box a "person" attribute label. By combining these pseudo-labels with the WSPD labels and pre-training them with the SSD, we were able to build a larger model to verify miss rates.

Furthermore, we pre-trained the constructed STPD and compared its detection performance with the model pre-trained using the WSPD. In order to examine the disparity in the miss rate among age attributes, it is essential to add an age attribute to the bounding box. Then, in order to evaluate the miss rate for each age attribute, we assigned "adult" and "children" labels to the INRIA Person Dataset, which is commonly used for person detection, using the models pre-trained with the WSPD and STPD, respectively. We also re-annotated the location of the bounding box. These two age attributes follow the age categories defined by the Statistics Bureau of the Ministry of Internal Affairs and Communications in Japan for (i) children (0–14 years) and (ii) adults (15 years and older). As a result, we constructed a pedestrian detection dataset consisting of 902 images and 2993 bounding boxes for training and evaluation. We named this dataset the Fairness-Aware INRIA Person Dataset (FA-INRIA). An example of the annotations and the breakdown of the dataset attributes are shown in Figure 2 and Table 3, respectively.

Adults

Children

Figure 2. Examples of age attribute annotation in Fairness-Aware INRIA Person Dataset (FA-INRIA).

Table 3. The age attributes in the Fairness-Aware INRIA Person Dataset (FA-INRIA).

Age	Images	Bounding Boxes
Adult	870	2672
Children	151	321
All	902	2993

3.3. Experimental Settings

In this paper, we compared the results under the same pre-training conditions. The batch sizes for pre-training the SSD were set to 64, 128 and 256, the number of epochs was set to 10, and the learning rate was set to 0.0005. When we conducted fine-tuning with the FA-INRIA using the pre-trained models on each dataset, the batch size was set to 4, the number of iterations was set to 12,000, and the learning rate was set to 0.0005. Furthermore, the training and test datasets were used with the same configuration as the original INRIA Person Dataset. The experimental settings described below also conform to these conditions.

3.4. Evaluation Metric

We only used the miss rate as an evaluation metric to assess the detection performance for adults and children. In person detection, the relationship between the miss rate and false positives is often evaluated for each image. However, our goal is to detect all ground truth bounding boxes. Therefore, we calculated the miss rate by examining the breakdown of the age attributes of the bounding boxes that could not be detected. The miss rate M is derived by the following equation:

$$MR = 1 - Recall \tag{2}$$

In this paper, we calculated the standard deviation to represent the miss rate disparity among age attributes:

$$MR_s td = \frac{1}{n}\sum_{i=1}^{n}(MR_i - \bar{MR})^2, \tag{3}$$

where n refers to the number of classes of attributes, which in this study was two ("Adult" and "Children").

3.5. Results

Table 4 shows the miss rate in the FA-INRIA Person Dataset using each of the pre-trained models. Compared to the model pre-trained with the WSPD, the model pre-trained with STPD reduced the miss rate by up to 0.4% for adults and up to 3.9% for children. In the WSPD pre-trained model, the disparity between the miss rates of adults and children was a maximum standard deviation of 4.6% and a minimum of 3.1%. In contrast, the STPD pre-trained model had a maximum standard deviation of 2.9% and a minimum standard deviation of 2.1%.

Table 4. Detection performance comparisons for our FA-INRIA. We use standard deviation to describe the disparity in detection rates between attributes. It is clear that our approach reduces the miss rate for all attributes.

Dataset	Batch Size, Epochs	MR_{adult}	MR_{child}	MR_{std}
WSPD	64, 10	13.9	23.1	4.6
	128, 10	13.8	21.2	3.2
	256, 10	13.1	19.2	3.1
STPD (ours)	64, 10	13.8	19.2	2.7
	128, 10	13.4	19.2	2.9
	256, 10	13.1	17.3	2.1

Then, the detection results of fine-tuning with the FA-INRIA using the pre-trained detectors on each dataset are shown in Figure 3, illustrating that the STPD pre-trained model is able to detect people that the WSPD pre-trained model misses.

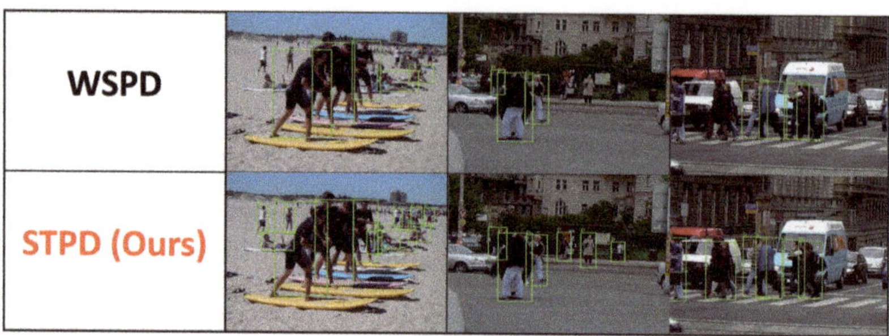

Figure 3. Comparison of detection results of WSPD and STPD.

4. Analysis and Discussion

4.1. The Relationship between the Bias in the Quantity of Data and the Miss Rate

In the aforementioned results, we successfully generated a pseudo bounding box containing a person from the Places365 dataset. In Figure 1, we present a visualization of the location of a person's bounding box that was predicted during the process of self-training. This method was implemented based on the success of self-training in object detection [14] and was found to reduce the miss rate for adults and children, respectively. Moreover, it is effective in collecting data on pedestrians regardless of their age attributes, and not only on children for whom the number of data is small. If the bias in the quantity of data between age attributes is the primary cause of the disparity in detection performance, then it is only the bounding boxes for children that need to be more efficiently collected. However, manual annotation is very costly and impractical. Therefore, we applied data augmentation to the children's bounding boxes in the FA-INRIA training data to investigate the effect on the miss rate for adults and children. In our work, we tried to augment the children's bounding boxes by applying horizontal flip.

Table 5 shows the detection performance when data augmentation is applied to the children's bounding boxes. It can be seen that when the batch size is 256, the miss rate for both attributes decreases. However, when the batch size is 64 or 128, the miss rate for children does not change, while the miss rate decreases for adults. These results indicate that applying data augmentation is effective in improving the overall detection performance. On the other hand, when we focus on the standard deviation, we must not forget that the disparity in detection performance between age attributes is expanding. First and foremost, a "person" can be an adult or a child. If the detection performance for adults is improved solely by increasing the data of children, we would consider that the bias in the quantity of data between classes is not directly relevant.

Table 5. The impact of applying data augmentation (horizontal flip) only to the bounding boxes of the children in the training data. The results show that applying data augmentation is effective in improving the overall detection performance. On the other hand, it may increase the disparity in detection performance among age attributes.

Batch Size,	MR_{adult}		MR_{child}		MR_{std}	
Epochs	w/o Aug.	w/ Aug.	w/o Aug.	w/ Aug.	w/o Aug.	w/ Aug.
64, 10	13.8	**12.6**	19.2	19.2	**2.7**	3.3
128, 10	13.4	**12.1**	19.2	19.2	**2.9**	3.6
256, 10	13.1	**10.7**	17.3	**15.4**	**2.1**	2.4

4.2. The Relationship between the Size of a Person's Bounding Box and the Miss Rate

Detecting small objects is a difficult task in object and person detection research because of the limited information that can be obtained from a bounding box with a small image size. It is clear that children have smaller bodies than adults. Therefore, the bounding boxes of children tend to be smaller than those of adults. Thus, we thought it would be important to investigate the size of bounding boxes in the FA-INRIA.

Figure 4 presents the distribution of the size of the bounding boxes for adults and children. Adults are shown in red and children are shown in blue. This distribution indicates that most of the bounding boxes that exceed the size of 600 pixels × 300 pixels in height and width, respectively, are for adults. In other words, the difference in the size distribution of the bounding boxes may be one of the factors affecting the disparity in the miss rate. Figure 5 also shows the distribution of the size of the bounding boxes in the image for the FA-INRIA (test set): the bounding boxes that could be detected are shown in red and the missed bounding boxes are shown in blue. As you can see in these figures, most of the missed bounding boxes are biased towards the smaller image size. In other words, in order to further mitigate the disparity in the miss rate, it is necessary to use detectors that can detect small persons.

In this paper, we investigated the effect of changing the image size of the input on the miss rate of each attribute. The SSD resizes the input image to a set size regardless of the size of the original image. This process is likely to result in the missing details of the image. In order to detect small bounding boxes, we thought that increasing the size of the input image would suppress the missing information. We examined three patterns of input image sizes: (i) 150 pixels × 150 pixels; (ii) 300 pixels × 300 pixels; and (iii) 600 pixels × 600 pixels. The default size for the SSD is 300 pixels × 300 pixels. For more accurate validation, we also used a sub-dataset with the same number of bounding boxes for adults and children in the training data.

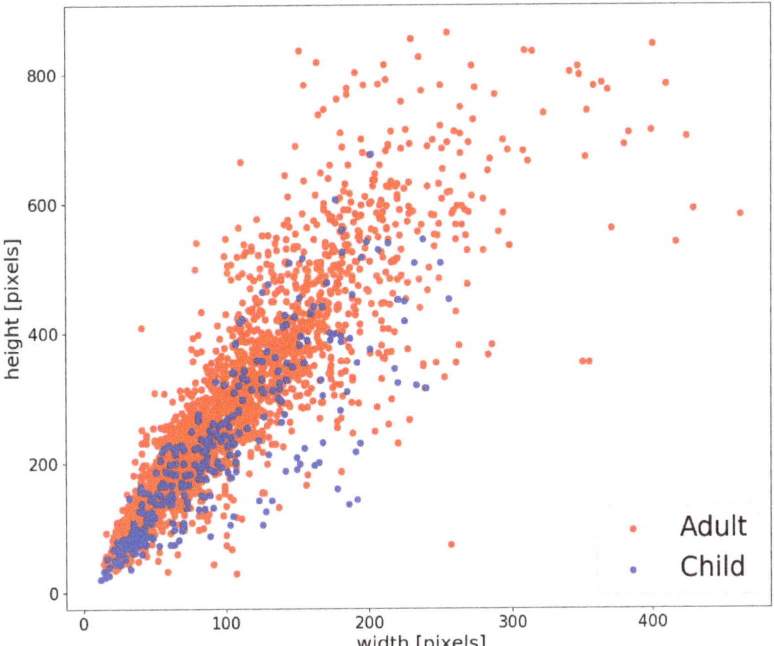

Figure 4. Distribution of bounding boxes for adults and children in the FA-INRIA Person Dataset. Children's bounding boxes tend to be relatively smaller than those of adults.

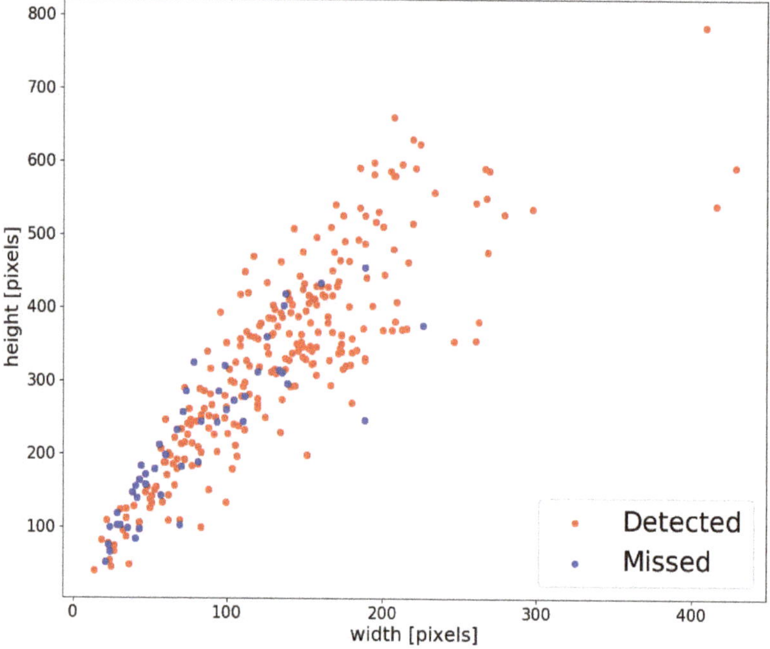

Figure 5. Whether bounding boxes can be detected in test data (red: detected, blue: missed).

Table 6 shows the miss rate when the input image size of the SSD is changed. It can be seen that increasing the size of the input image is a major factor in reducing the miss rate. On the other hand, when the input size is small (150 pixels × 150 pixels), the miss rate for children is very poor. We consider that this is because image information is also missing due to the relatively smaller bounding box. As shown in Figure 4, children's bounding boxes are more difficult to detect when the input size is small because children have a relatively higher proportion of small bounding boxes than adults. Based on this result and Figures 4 and 5, we conclude that the unbalanced distribution of the bounding box sizes is one of the main reasons for the disparity in detection performance between adults and children.

Table 6. The effect of changing the input size of the image to the SSD on the detection performance for each age attribute. The results show that increasing the input size decreases the miss rate. In addition, children are more strongly affected by changes in the size of the input. We conclude that the bias in the size of the bounding box is a major factor in the disparity in detection performance.

Batch	MR_{adult}			MR_{child}			MR_{std}		
Size,	Input size of the image (pixels × pixels)								
Epochs	150	300	600	150	300	600	150	300	600
64, 10	14.9	**14.1**	14.4	17.3	**15.4**	**15.4**	1.2	0.7	**0.5**
128, 10	15.2	14.6	**13.9**	21.2	17.3	**15.4**	3.0	1.4	**0.8**
256, 10	14.1	14.7	**13.4**	21.2	17.3	**15.4**	3.6	1.3	**1.0**

4.3. Appearance Difference

We considered two aspects: the bias in the quantity of data between classes and the size of the bounding boxes. However, as shown in Figure 5, we can see that some people are not detected even though the bounding box is relatively large. Moreover, as mentioned in Section 4.1, we found that the bias in the quantity of data between classes is most likely not relevant. These results suggest that there might be other factors that generate disparities in detection performance between age attributes. Subsequently, we hypothesized that there would be apparent differences between the distributions of bounding box sizes of adults and children as they differ significantly in size.

Figure 6 shows the compression of the image features using t-SNE and the visualization of the distribution. It is difficult to imagine that there is a disparity in detection performance based on the appearance of the distribution which is not clearly divided by age attribute and is evenly distributed. This result supports the fact that applying data augmentation to the children's bounding boxes was more effective in improving the detection rate for adults than for children. Since there is no apparent difference between adults and children, we reiterate that we do not need to consider the bias in the quantity of data between classes to reduce the miss rate for children.

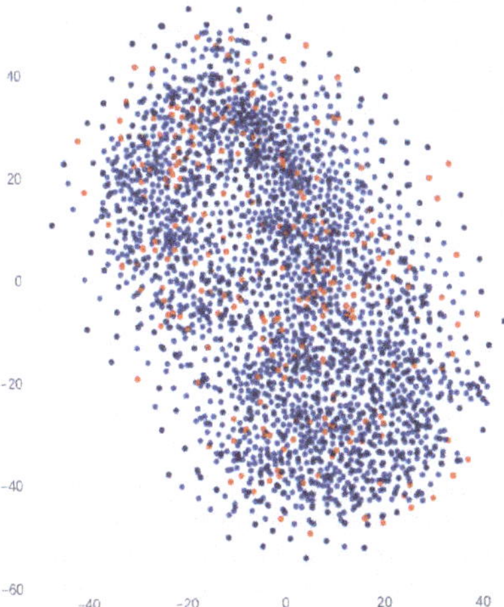

Figure 6. Data visualization of bounding boxes using t-SNE (blue: adults, red: children). There is no apparent significant difference between the bounding boxes of children and adults. As mentioned in Section 4.1, when data augmentation was applied to children's bounding boxes, the miss rate was strongly affected for adults but not for children. This data visualization supports the consideration that the bias in the quantity of data between classes has little to do with the disparity in detection performance.

5. Conclusions

In this paper, we investigated and examined various perspectives on the causes of disparity in the detection performance between adults and children in the task of pedestrian detection. As a first experiment, we confirmed that self-training extends the pre-training model and improves the overall detection performance. Then, we found that applying data augmentation to the bounding boxes of children—for whom there is less data available than for adults—significantly improves the detection performance for adults but not children. We also visualized the feature distribution of the bounding boxes using t-SNE and found that there was no apparent difference between adults and children. These results indicate that it is not necessary to consider the bias in the quantity of data in terms of age attributes in pedestrian detection.

On the other hand, when we looked at the size of the bounding boxes in our FA-INRIA, we observed that the distribution was biased toward a smaller size for children than for adults. In addition, we found that changing the input size of the image fed to the detector had a significant impact on detection performance for children. In other words, we concluded that the disparity in the size of the bounding boxes was a major factor in the disparity in detection performance among age attributes. In the future, focusing on the detection of small bounding boxes will help mitigate the bias between attributes.

Author Contributions: Main contribution: S.K., K.W. and R.Y.; supervision: Y.A., A.N. and H.K. All authors have read and agreed to the published version of the manuscript.

Funding: This research received no external funding.

Institutional Review Board Statement: Not applicable.

Informed Consent Statement: Not applicable.

Data Availability Statement: Not applicable.

Conflicts of Interest: The authors declare no conflict of interest.

References

1. Brandao, M. Age and Gender Bias in Pedestrian Detection Algorithms. *arXiv* **2019**, arXiv:1906.10490.
2. Minoguchi, M.; Okayama, K.; Satoh, Y.; Kataoka, H. Weakly Supervised Dataset Collection for Robust Person Detection. *arXiv* **2020**, arXiv:2003.12263.
3. Dalal, N.; Triggs, B. Histograms of oriented gradients for human detection. In Proceedings of the 2005 IEEE Computer Society Conference on Computer Vision and Pattern Recognition (CVPR'05), San Diego, CA, USA, 20–25 June 2005.
4. Girshick, R.; Donahue, J.; Darrell, T.; Malik, J. Rich Feature Hierarchies for Accurate Object Detection and Semantic Segmentation. In Proceedings of the IEEE Conference on Computer Vision and Pattern Recognition (CVPR), Columbus, OH, USA, 23–28 June 2014.
5. Girshick, R. Fast R-CNN. In Proceedings of the IEEE International Conference on Computer Vision (ICCV), Santiago, Chile, 7–13 December 2015.
6. Ren, S.; He, K.; Girshick, R.; Sun, J. Faster R-CNN: towards real-time object detection with region proposal networks. *IEEE Trans. Pattern Anal. Mach. Intell.* **2016**, *39*, 1137–1149. [CrossRef] [PubMed]
7. Liu, W.; Anguelov, D.; Erhan, D.; Szegedy, C.; Reed, S.; Fu, C.-Y.; Berg, A.C. SSD: Single Shot Multibox Detector. In Proceedings of the European Conference on Computer Vision (ECCV), Amsterdam, The Netherlands, 11–14 October 2016.
8. Redmon, J.; Divvala, S.; Girshick, R.; Farhadi, A. You Only Look Once: Unified, Real-Time Object Detection. In Proceedings of the IEEE Conference on Computer Vision and Pattern Recognition (CVPR), Las Vegas, NV, USA, 7–30 June 2016.
9. Zhao, Q.; Sheng, T.; Wang, Y.; Tang, Z.; Chen, Y.; Cai, L.; Ling, H. M2det: A single-shot Object Detector Based on Multi-level Feature Pyramid Network. *Proc. AAAI Conf. Artif. Intell.* **2019**, *33*, 9259–9266. [CrossRef]
10. Lin, T.-Y.; Goyal, P.; Girshick, R.; He, K.; and Dollar, P. Focal Loss for Dense Object Detection. In Proceedings of the IEEE International Conference on Computer Vision (ICCV), Venice, Italy, 22–29 October 2017.
11. He, K.; Gkioxari, G.; Dollar, P.; Girshick, R. Mask R-CNN. In Proceedings of the IEEE International Conference on Computer Vision (ICCV), Venice, Italy, 22–29 October 2017.
12. Wilson, B.; Hoffman, J.; Morgenstern, J. Predictive Inequity in Object Detection. *arXiv* **2019**, arXiv:1902.11097.
13. Zhou, B.; Lapedriza, A.; Khosla, A.; Oliva, A.; Torralba, A. Places: A 10 million Image Database for Scene Recognition. *IEEE Trans. Pattern Anal. Mach. Intell.* **2017**, *40*, 1452–1464. [CrossRef]
14. Zoph, B.; Ghiasi, G.; Lin, T.-Y.; Cui, Y.; Liu, H.; Cubuk, E.D.; Le, Q. Rethinking Pre-training and Self-training. *Adv. Neural Inf. Process. Syst.* **2020**, *33*, 3833–3845.

MDPI
St. Alban-Anlage 66
4052 Basel
Switzerland
Tel. +41 61 683 77 34
Fax +41 61 302 89 18
www.mdpi.com

MDPI Books Editorial Office
E-mail: books@mdpi.com
www.mdpi.com/books